Bamboo

Esther Titilayo Akinlabi · Kwame Anane-Fenin
Damenortey Richard Akwada

Bamboo

The Multipurpose Plant

Springer

Esther Titilayo Akinlabi
Department of Mechanical Engineering
Science
University of Johannesburg
Auckland Park
South Africa

Damenortey Richard Akwada
Department of Mechanical Engineering
Science
University of Johannesburg
Auckland Park
South Africa

Kwame Anane-Fenin
Department of Mechanical Engineering
Science
University of Johannesburg
Auckland Park
South Africa

ISBN 978-3-319-56807-2 ISBN 978-3-319-56808-9 (eBook)
DOI 10.1007/978-3-319-56808-9

Library of Congress Control Number: 2017943202

Printed on acid-free paper

This Springer imprint is published by Springer Nature
The registered company is Springer International Publishing AG
The registered company address is: Gewerbestrasse 11, 6330 Cham, Switzerland

Preface

Humans for many centuries have harnessed the multifaceted applications of bamboo, including those in the structural and engineering fields and for food and medicinal purposes mainly in Asia and Africa. However, the practical applications for bamboo in the areas of composite materials, textiles, and structural engineering have seen tremendous growth and research interest over the past decade.

The existing literature on bamboo often assumes a very narrow dimension. The literature on bamboo can be grouped into four main categories. These include biological, architectural, structural engineering, and the socio-economical contexts. Farrelly (1984), in "The Book of Bamboo: A comprehensive guide to this remarkable plant, its uses, and its history", focused on the biological components such as taxonomy, anatomy, and cultivation and harvesting. However, Janssen (2000) mainly dealt with the fundamental dimension of bamboo.

Bamboo is a multifunctional plant, and this book is structured to present a holistic perspective on bamboo by tackling all its relevant facets. It is however written from the perspective of the engineer. Bamboo used as a potential alternative to fossil fuels, forest firewood, and charcoal in the African context is discussed. This book will further present a thorough review of the research and literature, the current experimental characterisation techniques, and the role of finite element analysis in improving the properties of bamboo products. A comprehensive assessment of bamboo as an engineering and known engineering material is made. This book is intended for use by specialists, experts, and multidisciplinary research and development researchers working in fields related to bamboo. It would serve as a multidisciplinary research manual on bamboo taxonomy, properties, and, in particular, on bamboo characterisations and its industrial applications. More specifically, the authors' unique contributions to this book are in bamboo characterisation and the identification of different applications of bamboo as a sustainable engineering material, which has not previously been fully investigated and reported.

To the best of the authors' knowledge, no such book exists that presents a total view and also an African perspective on the subject. The African literature is critical because bamboo will assume a much greater role as the world gravitates towards sustainable resource utilisation. This book has been long overdue and would appeal to a wide demography due to its in-depth and broad spectrum while bridging the multidisciplinary divide to provide a holistic resource on the subject.

We would furthermore like to acknowledge the support of technical reviewers, as well as language and graphical editors, who have all contributed to this process. We value the system of scholarly peer-review and the approach that the same adds to producing a research text that adds to the body of scientific knowledge.

Johannesburg, South Africa
January 2017

Esther Titilayo Akinlabi
Kwame Anane-Fenin
Damenortey Richard Akwada

References

Janssen JJA (2000) Designing and building with Bamboo. International network for Bamboo and Rattan technical report No. 29, p 207
Farrelly D (1984) The Benefits of Bamboo. The Sciences 24: 11–12

Contents

Chapter 1
Bamboo Taxonomy and Distribution Across the Globe

Abstract Bamboo plants belong to the grass family and one of the subfamilies of Poaceae (Gramineae) and are one of the fastest-growing plants on the planet, with high rates of growth in tropical and subtropical climate regions. The bamboo plant or grass has cultural and ecological significance for many countries in Africa, America, and Asia, where it provides environmental, social, and economic benefits. Bamboos, therefore, grow in association with a wide variety of mostly mesic to wet forest types in both temperate and tropical regions, while some bamboos have adapted to more open grasslands or occur in more specialised habitats. Bamboo grows very fast, and, although the species type does have significant influence, generally, all bamboo matures quickly. This section of the book gives synoptic descriptions of this plant in respect to other grasses regarding its distribution across continents, morphology, and an updated phylogeny-based classification for the Bambusoideae. Emphasis is placed on an updated description of bamboo diversity regarding tribe and subtribe as well as the genera.

1.1 Introduction

The bamboo plant or grass has cultural and ecological significance in many countries within Africa, America, and Asia where it provides environmental, social, and economic benefits (Gratani et al. 2008; Oteng-Amoako et al. 2005). The distribution of different bamboo species across the tropical and subtropical regions naturally is dependent on certain conditions such as soil type, rainfall, temperature, and altitude. Bamboo grasses grow naturally in areas that receive an annual rainfall ranging from 1200 to 4000 mm, with average annual temperature ranging from 8 to 36 °C. They also grow well in different types of soils ranging from rich alluvium to hard lateritic soils, loamy soil, and sandy soils. The multi-purpose usage of bamboo grass taking into consideration its resilience, fast growth rates and the possibility of multiple harvesting within a few years are significant advantages over the other forestry species. The multipurpose use of bamboo grass as a substitute for timber has many benefits due to its lignified culms, fast growth, complex rhizome systems,

E.T. Akinlabi et al., *Bamboo*, DOI 10.1007/978-3-319-56808-9_1

and its sustainable and regenerative characteristics. The bamboo grass has a conservation value and can alleviate phenomena that emanate from global climate change (Lobovikov et al. 2007; Li and Shen 2011). Bamboo grass serves as an essential resource for many organisms, such as pandas and other living organisms. Bamboo is a subfamily of Poaceae (grass family) and comparatively like African napper grass, elephant grass, maize, wheat, millet, and sugar cane. It is an important grass inextricably linked to human livelihood, fulfilling the needs for shelter, food, clothing, and many others (Chaowana 2013; Wooldridge (2012). The range of its use is hardly rivalled in the plant kingdom, and it is not for nothing that bamboo is known as "the plant of multifunctional uses" (Sharma et al. 2014; Amada and Untao 2001; Lobovikov et al. 2007; Wooldridge (2012).

1.2 Bamboo's Identification

Bamboo is a versatile composite plant spanning different species. They are difficult to identify or classify. However, given their ecological and economic importance, correct identification is critical to their conservation, development, and a robust phylogenetic classification system underpins their identification (Judziewicz et al. 1999; Grass Phylogeny Working Group (GPWG) 2001; Bamboo Phylogeny Group (BPG) 2012; Kelchner and Bamboo Phylogeny Group 2013). There have been approximately 1100–1500 different species of bamboo which have been identified and classed into a range of over 115 genera species across the globe (Chaowana 2013; Steinfeld 2001; Nguyen 2006). Bamboo as a grass belongs to the group and order angiosperms and monocotyledon respectively. The bamboo grasses are a member of the subfamily Poaceae (Gramineae). They are categorised as grass but not as trees with twelve (12) large subfamilies namely Anomochlooideae, Arundinoideae, Aristidoideae, Danthonioideae, Ehrhartoideae, Micrairoideae, Pharoideae, Puelioideae, Pooideae, Chloridodeae, Panicoideae, and Bambusoideae grass (Sen and Reddy 2011; Yan et al. 2006).

Bamboo is divided into two major portions, the rhizomes and the culms (Steinfeld 2001; Tara Sen et al. 2011). The underground part of the culm (stem) is known as the rhizome and mostly sympodial or to a much lesser degree, monopodial. The rhizomes store the nutrients for growth and also secure the bamboo plant in the ground for its sustenance. The culm part is the upper portion of the bamboo grass that contains most of the woody material (Ahmed and Kamke 2005). The rhizomes consist of buds which develop into shoots that emerge from the ground to form a clump of culms. The culm of a bamboo is cylindrical and subdivided into sections by diaphragms or nodes. The section between two nodes along the culm are known as an internode (Amanda and Untao 2001). The culm of a bamboo by nature is hollow but has a thick wall which makes it ideal for the production of structural components. The bamboo culm is hollow with a diameter ranging from 0.25 to 12 in. (0.64–30.48 cm) and with a height from one foot to 120 feet (Amanda and Untao 2001). However, bamboo does not have any bark but has a hard, smooth outer skin due to the presence of silica. The branch systems, sheath, foliage leaves, and flowers complement the culm of the bamboo. Bamboo species differ from one another in features, such as culm growth style, which may be mainly erect or erect with droopy tips, clambering or arched. Bamboo has a natural composite material of cellulose fibres immersed in a lignin matrix with an average tensile resistance of about 700 MPa. It has a higher tensile strength, strength-to-weight ratio, and specific load-bearing capacity compared to mild steel. However, less energy is required for its processing. (Janssen 2000; Li and Shen 2011).

1.3 Bamboo Compared with Trees (Wood)

Bamboo is one of the fastest-growing plants on the planet, with reported higher rates of growth in tropical and subtropical climate regions compared to timber (wood). Bamboo grows very fast, but mostly the species type also has a significant influence. All bamboo matures quickly and might have 40–50 stems in one clump, which adds 10–20 culms yearly (Gratani et al. 2008; Latif and Liese 2002) as compared to wood which does not add a single shoot. Bamboo can attain its maximum height within four to six months with 15–18 cm (5–6 in.) daily increments, while wood takes years. The fast growth characteristic of bamboo gives it an added advantage for its utilisation as it is a high-yield renewable resource, cheap and abundant across the globe for multiple industrial applications. There are several different characteristics between bamboo and wood. Stress distribution within the full length of bamboo is even because of the nonexistence of knots or rays. It is hard to join bamboo because it often has thin walls and can be considered as a hollow tube. The chemical extracts in bamboo are different from those in other woods and therefore provide a much better bond when glued (Janssen 1995). The dimensional characteristics of bamboo, such as diameter, thickness, and internodal length, are structurally macroscopically graded while the distribution of fibre has a microscopically graded architecture, which leads to much more desirable properties (Amada et al. 1997).

Some species thrive in temperate climates in Africa, Asia, and the Americas. The bamboo plant is said to be one of the oldest building and reinforcement materials, having been used since ages. They grow on plains, hills, and high-altitude mountainous regions, and in most kinds of soils, except alkaline soils, desert, and marsh. They can grow from sea level to as high as 300 m. Bamboo growth is suitable in well-drained, sandy-to–clay, loamy soil or soil derived from underlying rocks with a pH of 5.0–6.5 (Janssen 2000; Oteng-Amoako et al. 2004, 2005; Gyansah et al. 2011).

1.3.1 Differences in Characteristics of Bamboo and Trees

Rhizomes and roots

i. Trees do not have rhizomes but consist of only underground roots, while bamboo culms consist of both rhizomes and roots within its underground part.

Stems

i. Trees stems are mostly solid in nature, while bamboo stems are hollow and segmented.
ii. The stem of the tree has, as its hardest part, the central part, while bamboo has, as its hardest section, the stem which is the periphery of the culm.
iii. It can be said that bamboo stems have no bark covering as compared with the stems of trees.
iv. The stems of trees grow slowly in height and diameter for several seasons, while the culms of bamboo grow extremely fast in height, ranging up to 36 m within six months), reaching full height within one growing season.
v. Tree stems grow as independent entities, while bamboo stems grow in association from a network of rhizomes, such that each culm depends on the others in a clump.
vi. Harvesting the stem of a tree does not have any direct influence on the remaining trees within the plantation, while, in the case of harvesting bamboo stems, there is a direct effect on the rest of the stems in the clump community.
vii. There is radial (lateral) communication throughout the stems of trees, while in bamboo there is no radial (lateral) exchanges in the stems except at the nodes.

1.4 Bamboo Origin and Its Habitats

The bamboo plant thrives very well in tropical and subtropical regions. It belongs to the subfamily Bambusoideae within the grass family of Poaceae. It consists of both herbaceous and woody bamboo and grows mostly in the forest as a grass but not a tree. The bamboo species in this group share some common features such as rhizomatous habit, hollow segmented culms, and petiolate blades with tessellating venation. Bamboo grass flowers with three or more lodicules, usually with six stamens, and its fruit possesses a little embryo and linear hilum as stated by Soderstrom (1981) and GPWG (2001). In recent years, some of the species of bamboo have been able to grow in open, grassy, or shrubby habitats at high elevations in the tropics (Clark et al. 2007; Soderstrom and Ellis 1988; Judziewicz et al. 1999).

The bamboo plant is distributed globally on all the continents except Antarctica and Europe. However, some species had been introduced to Europe a few years ago.

Table 1.1 Numbers of species of woody bamboos occurring in the countries of the Asia-Pacific region, Africa, and the Americas

Country/territory	Number of naturally occurring species (species without documented distributions in other nations)
Asia-Pacific	
China	626 (583)
India	102 (40)
Japan	84 (75)
Myanmar	75 (30)
Vietnam	69 (38)
Indonesia	56 (29)
Malaysia	50 (26)
Thailand	36 (4)
Philippines	26 (14)
Nepal	25 (6)
Papua New Guinea	22 (15)
Bhutan	21 (1)
Bangladesh	18 (0)
Laos	13 (4)
Sri Lanka	11 (6)
Hong Kong	9 (3)
Brunel	6 (1)
Korea, Republic	6 (2)
Cambodia	4 (0)
Australia	3 (2)
Pakistan	3 (0)
Singapore	3 (0)
Korea, DPR	2 (0)
Russia (Sakhalin and Kuril Ilands)	1 (0)
Africa	
Madagascar	33 (32)
Tanzania	4 (1)
Malawi	3 (0)
Uganda	3 (0)
Zambia	3 (0)
Cameroon	2 (0)
Congo	2 (0)
DR Congo	2 (0)
Ethiopia	2 (0)
Sudan	2 (0)
Zimbabwe	2 (0)
Angola	1 (0)

(continued)

Table 1.1 (continued)

Country/territory	Number of naturally occurring species (species without documented distributions in other nations)
Benin	1 (0)
Burundi	1 (0)
The Central Africa Republic	1 (0)
Comoro Island	1 (0)
Cote d'Ivoire	1 (0)
Eritrea	1 (0)
Gambia	1 (0)
Ghana	1 (0)
Guinea	1 (0)
Guinea-Bissau	1 (0)
Kenya	1 (0)
Lesotho	1 (0)
Mozambique	1 (0)
Nigeria	1 (0)
Rwanda	1 (0)
Senegal	1 (0)
Sierra Leone	1 (0)
South Africa	1 (0)
Togo	1 (0)
North, Central, and South America	
Brazil	134 (110)
Venezuela	68 (32)
Colombia	56 (15)
Ecuador	41 (14)
Costa Rica	36 (13)
Peru	35 (15)
Mexico	32 (17)
Bolivia	20 (2)
Panama	19 (1)
Chile	14 (10)
Cuba	13 (6)
Argentina	12 (2)
Guatemala	12 (0)
Honduras	8 (0)
El Salvador	7 (1)
Haiti	7 (0)
Nicaragua	7 (0)
Trinidad and Tobago	7 (2)

(continued)

Table 1.1 (continued)

Country/territory	Number of naturally occurring species (species without documented distributions in other nations)
Dominican Republic	6 (0)
Paraguay	6 (1)
Guyana	5 (2)
Puerto Rica	5 (0)
Uruguay	5 (1)
Suriname	4 (0)
Belize	3 (1)
Bahamas	1 (0)
Dominica	1 (0)
French Guiana	1 (0)
Guadeloupe	1 (0)
Jamaica	1 (0)
Martinique	1 (1)
United States	1 (1)
Virgin Island	1 (0)

Source Ohrnberger (1999)

The major species are distributed in Asia (China with about 626 species, India with 102 species, Japan with 84, Myanmar 75, and Malaysia with 50 species). Also in South America, Brazil has 134 species, Venezuela 68, Colombia 56, while Africa has about five species, all of which are listed and described in Table 1.1 (Bystriakova et al. 2003a, b). It is distributed from latitude 47° N to 50°30′ N and grows at altitudes from sea level to 4300 m (Soderstrom and Calderon 1979; Judziewicz et al. 1999; Ohrnberger 1999). Bamboo occupies a broad range of grass habitat types, especially in forests, from temperate to tropical climatic zones where they are often dominant and are highly visible elements of the vegetation (BPG 2012).

1.4.1 Bamboo Habitations

The occupancy of bamboo plants in an ecological area includes temperate deciduous forests, coniferous forest, lowland tropical forests, mountainous forests, understory, (moist) wetter forests, grasslands, and many other regions. They are mostly distributed across the tropics and subtropics zones through natural occurrence, as well as cultivation on farmlands (Table 1.2).

Table 1.2 Bamboo and its Habitats

Source	Species	Habitat type	Countries
Stapleton (1994a, b); Li and Xue (1997), Taylor and Qin (1997), Triplett et al. (2006), Dai et al. (2011)	*Arundinarieae*	Temperate deciduous to coniferous forest or both	Eastern Asia and Eastern North America
Taylor and Qin (1997), Noguchi and Yoshida (2005), Tsuyama et al. (2011)	*Arundinarieae*	They are mostly distributed in the understory and dominant in the wetter ecology	Eastern Asia and Eastern North America
Judziewicz et al. (1999), Seethalshmi and Kumar (1998), BPG (2012)	*Bambuseae*	Lowland tropical forest	China, Thailand, Vietnam, India
Li and Xue (1997), Taylor and Qin (1997),	*Bashania, Chimonobambusa, Fargesia, Indosasa, Yushania*	Montane forests	China
Stapleton (1994a, b, c)	*Thamnocalamus and Drepanostachyum*	Seasonally dry	Central Himalayas
Stapleton (1994a, b, c)	*Yushania, Sarocalamus, Chimonobambusa*	Wetter side of mountain ranges	China
Noguchi and Yoshida (2005), Tsuyama et al. (2011)	*Sasa, Sasamorpha*	Wetter forests (Understory)	China, Korea, and Japan
Stapleton (1994a, b, c), Li and Xue (1997), BPG (2012).	*Acidosasa, Drepanostachyum, Indosasa, and Sinobambusa*	Dry or evergreen subtropical forests	Asia
Triplett et al. (2006)	*Arundinaria*	Woodlands and forests, mostly along rivers	Eastern USA
Judziewicz et al. (1999), Triplett et al. (2006)	*Switch cane (A. tecta)*	Swamps	Southeastern USA
Yang and Xue (1990), Dai et al. (2011)	*Arundinaria*	Temperate bamboo-dominated	China, Japan, North America
Soderstrom and Calderón (1979), Seethalakshmi and Kumar (1998), Judziewicz et al. (1999), BPG (2012)	*Bambusa, Dendrocalamus, Eremocaulon, Guadua, Gigantochloa, and Schizostachyum*	Lowland moist tropical forests or lower montane forests	China, Mexico
Soderstrom and Londoño (1988), Dransfield (1992, 1994)	*Alvimia, Chusquea, Dinochloa, Hickelia, Neomicrocalamus, and Racemobambos*	Lowland tropical bamboos	America, Asia

(continued)

Table 1.2 (continued)

Source	Species	Habitat type	Countries
Seethalakshmi and Kumar (1998), BPG (2012), Gopakumar and Motwani (2013)	*Ochlandra*	Reed-like thickets along stream banks	Asia, Africa, India, Australia, Sri Lanka
Soderstrom and Calderón (1979), Gadgil and Prasad (1984), Rao and Ramakrishnan (1988), Seethalakshmi and Kumar (1998), Ruiz-Sanchez et al. (2011)	*Dendrocalamus strictus, Guadua paniculata, Otatea Perrierbambus*	Lowland species or genera and drier forest types	India, Latin America, Mexico and Colombia, Madagascar
Soderstrom and Calderón (1979), Judziewicz et al. (1999)	*Actinocladum and Filgueirasia*	Lowland bamboos	Brazil
Stapleton (1998)	*Vietnamosasa*	Grassland	Indochina
Judziewicz et al. (1999)	*Guadua*	Natural tropical bamboo forests	Amazon Basin
Judziewicz et al. (1999)	*Aulonemia, Chusquea, Rhipidocladum*	Andean montane forests	Brazil
Judziewicz et al. (1999)	*Chusquea and Merostachys*	Atlantic forests	Brazil
Judziewicz et al. (1999)	*Chusquea*	Cloud forest and pine–oak–fir forests	Mexico, Central and South America
Judziewicz et al. (1999), BPG (2012)	*Chusquea*	Nothofagus or Araucaria forests	Chile and Argentina
Judziewicz et al. (1999), BPG (2012)	*Nastus borbonicus, Chusquea*	Montane forests	Chile and Argentina
Dransfield (1992), Wong (1993)	*Holttumochloa, Racemobambos*	Montane forests	Southeast Asia
Soderstrom and Calderón (1979), Soderstrom and Ellis (1982), Judziewicz et al. (1999)	*Arundinarieae and Bambuseae*	High elevation grasslands or shrublands	China, Eastern and Southwest Asia, Japan, Madagascar, Sri Lanka
Li et al. (2006), BPG (2012)	*Fargesia (Arundinarieae), Yushania (Arundinarieae)*	Temperate mountains	China, Eastern and Southwest Asia, Japan, Madagascar, Sri Lanka
Soderstrom and Ellis (1982, 1988)	*Arundinarieae, Bergbambos* Kuruna	Tropical mountain grasslands and shrublands	South Africa, Sri Lanka and India respectively

(continued)

Table 1.2 (continued)

Source	Species	Habitat type	Countries
Judziewicz et al. (1999), Viana et al. (2013)	*Aulonemia, Cambajuva*	Mono-dominant stands at high elevations	Brazil
Li et al. (2006)	*Arundinarieae, Fargesia yulongshanensis*	Might elevations	Chinese or Himalayan
Judziewicz et al. (1999), Fisher et al. (2009), BPG (2012)	*Chusquea acuminatissima, Chusquea aristata, Chusquea tessellate, Chusquea aristata, Chusquea guirigayensis, villosa, Chusquea tessellate, Chusquea*	Highest elevational (4000–4400 m)	Central and South America, Mexico,
Judziewicz et al. (1999)	*Pariana,*	lowland tropical forests	Atlantic Brazil, Amazonian Bolivia, Costa Rica
Judziewicz et al. (1999), Judziewicz and Clark (2007)	*Cryptochloa, Lithachne, Pariana, Raddiella (R. esenbeckii), and Olyra (O. latifolia)*	lower montane forests	Agentina, Brazil, America, Mexico
BPG (2012), Ferreira et al. (2013)	*Ekmanochloa*	savannas or wet cliff faces (semi-deciduous seasonal forests)	Argentina, Brazil, America, Mexico, West Indies
Judziewicz et al. (1999), Judziewicz and Clark (2007), BPG (2012)	*Lithachne or Olyra*	minimal diversity near the equator	Brazil, Colombia, Cuba, Panama
Judziewicz et al. (1999)	*Reitzia*	Atlantic forests	Brazil
Soderstrom and Calderón (1979), Soderstrom et al. (1988), Ferreira et al. (2013)	*Olyreae (herbaceous)*	Atlantic forests	American, Brazil, Papua New Guinea, Cuba, Colombia, Panama

1.5 Molecular Taxonomy of Bamboos

The journey to derive taxonomical data for classification purposes has been a difficult one since little to no data existed on the morphology of bamboo. The last decade of the 20th century has seen a much more scientific and objective approach

Table 1.3 Fundamental studies on molecular taxonomy of bamboo

Researcher	Method	Notes/Findings
Friar and Kochert (1991), Taguchi-Shiobara et al. (1998)	DNA fragment using restriction fragment length polymorphism (RFLP)	Studies were conducted on Phyllostachys
Watanabe et al. (1994)	DNA fragment using RFLP of chloroplast DNA	Asian bamboos were used for studies. Investigations led to the recovery of a subtribe clad that represents Bambusinae sensu which includes Bambusa, Dendrocalamus, Gigantochloa and Thyrsostachys
Gielis et al. (1997)	Random amplified polymorphic DNA(RAPD) markers were employed	Studies were conducted on Phyllostachys
Loh et al. (2000)	Amplified fragment length polymorphism (AFLP) markers and RAPD were used	A close relationship was established between Bambusa and Gigantochloa
Pattanaik (2008)	Amplified fragment length polymorphism (AFLP) markers were used	The closeness of Dendrocalamus sensulato to Bambusa balcooa was determined by the placement of the latter within the former. The recovery of Melocalamus and Thyrsostachys as sister lineages to Dendrocalamus and Bambusa was achieved
Kelchner and Clark (1997)	DNA based on sequence using chloroplast rpl16 intron sequences is employed	The phylogenetic reconstruction in Chusquea was achieved
Guo et al. (2001, 2002), Guo and Li (2004)	DNA sequence-based research using an internal transcribed spacer (ITS) and the “Granule-Bound Starch Synthase I” (GBSSI) gene sequence	The samples used were in the group Thamnocalamus and its related allies
Sun et al. (2005)	DNA sequence-based research using Internal transcribed spacer (ITS) sequence	The samples used were in Bambusa group
Hodkinson et al. (2000)	AFLP and ITS approaches were used simultaneously	Reconstruction of the phylogenetics of Phyllostachys is achieved
Sungkaew et al. (2009)	Recircumscription using molecular data	Evidence from molecular data was used to establish that Bambusoideae has three (3) tribes namely Bambuseae, Arundinarieae, and Olyreae)

Source Pattanaik and Hall (2011)

to accessing, reviewing, and analysing molecular phylogenetic information of the family Poaceae (Pattanaik and Hall 2011). Previously, taxonomists utilised extractions such as isozymes and phenolics from the deoxyribonucleic acid (DNA) of bamboo species to establish inter and intra connections and relationships between variables such as polymorphic valuation, taxa, and identification of species (Chou and Hwang 1985; Alam et al. 1997). In recent times, however, taxonomists have investigated DNA and successfully established variations within it. During investigations, two types of data were generated for analysis, namely gene sequence and restriction fragmented data. The analysis was conducted using the maximum likelihood and maximum parsimony mathematical models (Pattanaik and Hall 2011). In the study of Phyllostachys for example, DNA fragments were derived using restraint fragmentation of length polymorphism (RFLP) (Friar and Kochert 1991; Taguchi-Shiobara et al. 1998).

Fundamental studies conducted on the molecular taxonomy of bamboo are summarised in Table 1.3.

The current research trends show that Bambusoideae receives considerable attention regarding molecular taxonomy analysis (BPG 2012). This is evident in the role the Bamboo Phylogeny Group is playing to ensure that a globally robust and molecular data-based phylogeny for classification Bambusoideae is established. This process started in 2005.

1.6 Scientific Classifications and Characteristics of Bamboos

Bamboos are classified into three tribes based on their molecular and morphological characters. The molecular and morphological characters include Arundinarieae, which is a woody bamboo and Bambuseae which is also a tropical bamboo, as well as Olyreae, which is a herbaceous bamboo (Sungkaew et al. 2009; BPG 2012; Kelchner and Bamboo Phylogeny Group 2013). Bambusoideae (bamboos) is one of the twelve (12) currently recognised subfamilies of Poaceae (grasses) (GPWG 2001; GPWG II 2012). It has received much attention via comprehensive molecular analyses in the bamboo family (GPWG 2001; Wu and Ge 2012; Bouchenak-Khelladi et al. 2008; GPWG II 2012). Bamboo morphology is complicated in nature with some species possessing sporadic flowering features that affect the vegetation. However, they are preserved in their habitat having broad pseudopetiole leaves with fusoid cells in the mesophyll (Judziewicz et al. 1999; Clark 1997). Bamboos are the only diversified grass in the global forests (Sungkaew et al. 2009; Judziewicz and Clark 2007; Zhang and Clark 2000).

A study by BPG (2012) and GPWG (2001) shows that there are 1439 described species in 115 genera. These tribes, including their species, consist of Arundinarieae, which is a temperate woody bamboo having some 533, Bambuseae, a tropical woody bamboo having 784, and Olyreae, herbaceous bamboos with 122.

However, there are still new genera and species being further discovered in the three (3) main tribes and being described with phylogenetic analysis supporting the circumscriptions (Fisher et al. 2009).

There are two clades within the Bambuseae comprising paleotropical woody bamboos and Neotropical woody bamboos. A Neotropical woody bamboo clade is subdivided into three (3) subtribes including Guaduinae, Arthrostylidiinae, and Chusqueinae, while the Paleotropical woody bamboo clade is currently composed of the four subtribes: Melocanninae, Bambusinae, Racemobambosinae, and Hickeliinae. The Olyreae contains three subtribes: Buergersiochloinae, Olyrinae, and Parianinae (Sungkaew et al. 2009; BPG 2012; Kelchner and Bamboo Phylogeny Group 2013). Morphologically, the herbaceous bamboos can be distinguished from the woody bamboos because they are made up of only weakly lignified culms, no culm leaves, and no outer ligules, but these characteristics are found in the woody bamboos (BPG 2012). The unisexual spikelets and seasonal flowering in the herbaceous bamboos are different from that of bisexual spikelets with monocarpic flowering. The unisexual spikelets' flowers are gregarious at long intervals which are a valid synapomorphy in the woody bamboos. The olyroid silica body is present in the herbaceous bamboos (except for Buergersiochloa), but not found in woody bamboos. Moreover, Olyreae are diploid, whereas woody bamboos are polyploid. Additionally, the tropical woody bamboos are distinct from the temperate woody bamboos through their rhizome type and branch development (Zhang and Clark 2000; Judziewicz and Clark 2007; BPG 2012).

Bambusoideae is a member of the monophyletic group and sister to the Pooideae clade (GPWG 2001; Wu and Ge 2012; Triplett and Clark (2006); Zhang et al. 2012; BPG 2012). Within the Bambusoideae, each tribe is also supported as a monophyletic group with high confidence (Sungkaew et al. 2009; Triplett and Clark (2006); Kelchner et al. 2013).

The woody bamboos consist of culm leaves (leaves modified for the protection and support of the tender young shoots). The woody bamboo is made up of complex vegetative branching, with an outer ligule (contra ligule) on the foliage leaves and gregarious monocarpy (GPWG 2001; Judziewicz and Clark 2007). The flowering cycle of bamboo ranges between 30 and 120 years for a bisexual flower (GPWG 2001; Judziewicz et al. 1999; Judziewicz and Clark 2007). The herbaceous bamboo lacks differentiated culm leaves, outer ligules and has restricted vegetative branches as compared to woody bamboos. Herbaceous bamboo flowers seasonally and a unisexual spikelet (Judziewicz et al. 1999; Judziewicz and Clark 2007) Olyreae bamboos on the other hand, excluding New Guinea's endemic Buergersoichloa, which have crenate silica bodies, do not flower seasonally and do not have a unisexual spikelet (Zhang and Clark 2000; Clark et al. 2007).

Woody bamboos, which colonise forest gaps and edges, are an essential element in tropical and temperate forests, especially in mountainous regions (Li and Xue 1997; Judziewicz et al. 1999; Zhou et al. 2011). These species of bamboos may be canopy or understory dominant, and they may form nearly mono-dominant stands that can cover extensive areas (Wong 1991, 1995, 2004; Judziewicz et al. 1999). Woody bamboos are important and sometimes dominant in high-altitude grasslands in both tropical and sub-temperate montane systems (Soderstrom et al. 1988; Judziewicz and Clark 2007). Woody bamboos have a colonising ability as understory dominants, and their characteristics play a major role in the ebb and flow of forest dynamics. The gregarious, monocarpic flowering of woody bamboos permit the forest to reoccupy previously disturbed sites (Wong 1991; Judziewicz et al. 1999; Martins et al. 2004). The effects of bamboo flowering and the subsequent death of woody bamboos affect the ecology of the vegetation. The effects contribute to soil fertility, water retention, and nutrient uptake in the vegetation (Marchesini et al. 2009). These effects of bamboo also affect the regeneration of other forest species in various ways (Larpkern et al. 2010). Carbon sequestration is environmentally effective in bamboo plantations or forests across the globe (Zhou et al. 2011).

Historically, there has been a comprehensive analysis of the Bambusoideae subfamily with extensive samplings. Although much is still to be achieved, researchers have derived many results through standard phylogenetic analyses of diverse lineages, grounded on both plastid and nuclear data (Table 1.4). The table shows comparisons of the historical classifications of Bambusoideae. There have been some revisions and restructuring in the current compilation and review of the phylogenetic findings on the classification of tribal and subtribal bamboos; this is a result attained based on a synthesis. The BPG team are in the process of preparing a separate phylogenetic analysis of plastid orders for the Bambusoideae tribes and subtribes.

1.6.1 Diversification of Bamboo

The bamboo subfamily (Bambusoideae) traditionally included what is now recognised as the "true" bamboos as well as some other taxa now classified as subfamilies or tribes within other subfamilies of grasses (Clayton and Renvoize 1986; Soderstrom and Ellis 1987). From this subfamily, some molecular analysis which is unequivocal is demonstrated using the polyphyly of the traditionally circumscribed Bambusoideae (Clark et al. 1995; Zhang and Clark 2000; GPWG 2001). The narrower circumscription of the subfamily is currently accepted, and consistently it has been supported by all the analyses, indicating that they have adequate taxon sampling (Sungkaew et al. 2009; GPWG II 2012). The analysis of Kelchner and Bamboo Phylogeny Group (2013) could be relevant in this case as he indicates a likelihood of woody bamboo possessing a single lineage based on the chloroplast sequence data stating that the herbaceous bamboos are resolved as sisters to the tropical woody, bamboos with the temperate woody bamboos as sisters to that clade, respectively. Additionally, Kelchner et al. (2013) in the findings also supports the relationships within Olyreae in which Buergersiochloinae, represented by Buergersiochloa Bambusoideaes, was sister to the remaining herbaceous bamboos.

Table 1.4 Comparison of historical classifications of morphological phylogenetic analyses on Bambusoideae

Authors	Year	#Taxa (Focus)	Data Used	Findings
Friar and Kochert	1994	1 g, 20 spp. *(Phyllostachys)*	nuclear RFLPs	Support for the division of *Phyllostachys* into two broad groups; the utility of RFLPs for species-level problems
Gielis et al.	1997	1 g, 42 spp. *(Phyllostachys)*	RAPDs	Support for *Phyllostachys* section *Heteroclada*; utility and reliability of RAPDs in identifying genotypes within *Phyllostachys*
Guala et al.	2000	19 g, 21 spp. (woody bamboos)	*ndhF* (3' end)	Polyphyly of *Apoclada* confirmed, with *A. simplex* (type species) in the Guaduinae and the other two species forming a clade in the Arthrostylidiinae
Hodkinson et al.	2000	4 g, 15 spp./1 g, 23 spp. (*Phyllostachys*)	ITS, AFLPs	Support for monophyly of *Phyllostachys* and its main subgeneric groups; the utility of AFLPs for phylogenetic reconstruction among closely related species
Guo et al.	2001	3 g, 23 spp. (temperate woody)	ITS	Non-monophyly of *Yushania*, *Fargesia*; support for some species groupings within alpine bamboos
Guo et al.	2002	7 g, 31 spp. (temperate woody)	ITS	Putative support for monophyly of the *Thamnocalamus* group of the temperate woody bamboos and position of *Chimonocalamus* as sister to the remainder of this group; support for monophyly of *Ampelocalamus*
Nayak et al.	2003	6 g, 12 spp.	RAPDs	Two clusters are recovered but no correspondence to taxonomic groupings

(continued)

Table 1.4 (continued)

Authors	Year	#Taxa (Focus)	Data Used	Findings
Guo and Li	2004	8 g, 31/33 spp. (temperate woody)	ITS, GBSSI	Support for monophyly of *Ampelocalamus* and *Chimonocalamus*; polyphyly of *Thamnocalamus* and the *Thamnocalamus group*
Sun et al.	2005	3 g, 21 spp.	ITS	Polyphyly of *Bambusa*
Zhuge et al.	2005	3 g, 17 spp. (*Arundinaria* and related genera)	ITS, trnL-trnF	Moderate support for what are now recognised as the *Arundinaria* and *Phyllostachys* clades within the temperate woody bamboos
Clark et al.	2007	27 g, 46 spp. (*Chusqueinae*, *Hickeliinae*)	*rpl16* intron, morphology	Support for monophyly of Bambusoideae, Bambuseae (moderate), Olyreae, Chusqueinae, and Madagascan Hickeliinae; possible paraphyly of Neurolepis; aspects of morphological evolution
Das et al.	2007	4 g, 15 spp. (*Bambusa, Dendrocalamus, Pseudobambusa, Gigantochloa*)	RAPDs, vegetative morphology	Relationships among the 15 species, based on allelic polymorphism data consistent with Gamble (1896); demonstration of the potential use of RAPDs for evaluation of phylogenetic relationships
Peng et al.	2008	25 g, 43 spp. (temperate woody)	ITS, GBSSI	Support for monophyly of the temperate woody bamboos; non-monophyly of traditional subtribes and many genera; support for inclusion of *Menstruocalamus* and *Qiongzhuea* in *Chimonobambusa*

(continued)

Table 1.4 (continued)

Authors	Year	#Taxa (Focus)	Data Used	Findings
Sharma et al.	2008	6 g, 21 spp. (1° Bambusinae + *Phyllostachys*, *Sasa)*	SSRs	Support for monophyly of *Phyllostachys* and groupings of *Bambusa* + *Dendrocalamus* and *Melocanna* + *Ochlandra*; rice genomic SSRs and sugarcane ESTSSRs can be transferable to bamboo (to 44.8 and 75% respectively)
Sungkaew et al.	2009	33 g, 52 spp. (Bambusoideae)	*matK, rps16* intron, *trnL* spacer, *trnL-F* spacer, *atpBrbcL* spacer	Confirmed paraphyly of woody bamboos; Melocanninae as sister to the remaining paleotropical woody bamboos; recognition of three tribes: Arundinarieae, Bambuseae, Olyreae
Goh et al.	2010	9 g, 24 pp (Bambusinae)	GBSSI, *rps16- trnQ, trnCrpoB, trnH-psbA and trnD-T*	Non-monophyly of *Bambusa*; climbing Southeast Asian genera are distinct from the core *Bambusa* group
Zeng and Zhang	2010	26 g, 148 spp. (temperate woody)	*atpI-H, psaAORF170, rpl32- trnL, rpoB-trnC, rps16-trnQ, trnD-T, trnS-G, trnT-L*	Confirmed monophyly of Arundinarieae and polyphyly of traditional subtribes; support for ten major clades within the tribe; low molecular divergence within the tribe
Pattanaik and Hall	2011	1 g, ten spp. (Dendrocalamus)	AFLPs	Evidence for polyphyly of Dendrocalamus; there is no support for previous infrageneric classifications of the genus
Wu and Ge	2012	3 g, three spp. Bambusoideae, 19 spp. other kinds of grass	76 chloroplast protein encoding genes	Support for (Bambusoideae + Pooideae) Ehrhartoideae

Adapted from BPG (2012); GPWG (2001); Judziewicz and Clark (2007)

1.6.2 Bamboos Tribes, Subtribes, and Genera Classification

Table 1.5 presents categories of group classification of Bambusoideae into tribes, subtribes, and genera. This outlines the various class of tribes of bamboos with their subtribes as well as the genera in the subfamily (BPG 2012).

1.7 Important Bamboo Genera and Species

1.7.1 Important Bamboo Genera

There are over 115 genera of woody bamboos in the world according to a report by BPG (2012) and GPWG (2001). They are native to Africa, the Americas, Asia, and Oceania. Some significant genera have been identified by describing their characteristics. Importantly, it should be noted that a bamboo genus poses the features of being sympodial and does form discrete clumps. Table 1.6 give some features of selected genera of bamboo.

1.7.2 Important Bamboo Species

There are over 1439 species of woody bamboos in the world, according to a report by BPG (2012) and GPWG (2001). They are native to Africa, the Americas, Asia, and Oceania, and some species have been introduced into Europe in recent times. Some important bamboo species have been identified and their characteristics described. Importantly, it should be noted that all bamboo species possess the characteristics of being sympodial and form discrete clumps. Table 1.7 illustrates important bamboo species and their characteristics.

1.7.3 Other Less/Known Taxa of Value

Arundinaria alpina: This species of bamboo popularly known as A. alpina or 'highland bamboo' is distributed over a vast area of forests in Ethiopia and its neighbouring countries, including Burundi, Kenya, Rwanda, Uganda, and other African nations. This species is seen as being associated mostly with African nations, and the locals mostly use it for general purposes including scaffoldings, walling or fencing, and other agricultural works.

Table 1.5 Bamboo Tribe, Subtribe, and Genera of Bambusoideae

Tribe	Subtribe	Genera
Arudinarieae (Temperate woody bamboo) This tribe of bamboo comprise about 30 genera with about 546 species. This tribe of bamboos are said to have the longest known flowering cycles of up to 120 years (Judziewiez and Clark 2007)		*Acidosasa, Ampelocalamus, Arundinaria, Bashania, Bergbamboo, Borinda, Chimonobambusa, Chimonocalamus, Drepanostachyum (Himalayacalamus), Fargesia, Ferrocalamus, Gaoligongshunia, Gelidocalamus, Himalayacalamus, Indocalamus, Indosasa, Kuruna, Oldeania. Oligostachyum, Pleioblastus, Pseudosasa, Sarocalamus, Sasa, Sasaella, Sasamorpha, Semiarundinaria, Shibataea, Sinobambusa, Thamnocalamus, Yushania*
Bambuseae (Tropical woody bamboo): This tribe of bamboo comprise of eight (8) subtribes with about 70 genera as well as 812 species. Bambuseae is distinctly classified into two main groups, namely; Neotropical, and Paleotropical and woody bamboos with its own subtribes and species given respectively (Clark et al. 2007; Ramanayake et al. 2007; BPG 2012)	1. Arthrostylidiinae	*Actinocladum, Alvimia, Apoclada, Arthrosstylidium, Athroostachys, Atractantha, Aulonemia (Matudacalamus), Cambajuva, Colanthelia, Didymogonyx, Elytrostachys, Filgueirasia, Glaziophyton, Merostachys, Myriocladus, Rhipidocladum*
	2. Chusqueinae	*Chusquea, Neurolepis*
	3. Guaduinae	*Apoclada, Eremocaulon, Guadua, Olmeca, Otatea.*
	4. Bambusinae	*Bambusa, Bonia (Monocladus), Cyrtochloa, Dendrocalamus (Klemachloa, Oreobambos, or Sinocalamus) Dinochloa, Fimbribambus, Gigantochloa, Greslania, Holttumochloa,*

(continued)

Table 1.5 (continued)

Tribe	Subtribe	Genera
		Kinabaluchloa, Maclurochloa, Melocalamus, Mullerochloa, Neololeba, Neomicrocalamus. Oreobambos, Oxytenanthera, Parabambus, Pinga, Pseudobambus, Pseudoxytenathera, Soejatmia, Sphaebambos, Temochloa, Temburongia, Thyrsostachys, Vietnamosasa
	5. Hickeliinae	*Cathariostachys, Decaryochloa, Hickelia, Hitchcockella, Nastus, Perrierbambus, Sirochloa, Valiha.*
	6. Melocaninnae	*Cephalostachyum, Davidsea, Leptocanna, Melocanna, Neohouzeaua, Ochlandra, Pseudostachyum, Schizostachyum, Staplatonia, Teinostachyum.*
	7. Racemobambodinae	*Racemobambos*
	8. Shibataeinae	*Qiongzhuea, (Brachystachyum), Temburongia (incertae sedis)*
Olyreae (Herbaceous bamboo): This tribe of bamboo are classified into three (3) subtribes based primarily on morphological data with 22 genera made up of 124 species. (Soderstrom and Ellis 1987; Judziewicz and Clark 2007). The Phylogenetic relationships within the Olyreae have not been examined extensively, although preliminary data support these three lineages and indicate the likely paraphyly of Pariana and probably polyphyly of Olyra (GPWG 2001; BPG 2012)	1. Buergersiochloineae	*Buergersiochloa*
	2. Parianinae	*Eremitis, Pariana, Parianella*
	3. Olyrineae	*Agnesia, Arberella, Cryptochloa, Diandrolyra, Ekmanochloa, Froesiochloa, Lithachne, Maclurolyra, Mniochloa, Olyra, Parodiolyra, Piresia, Piresiella, Raddia, Raddiella, Rehia, Reitzia, Sucrea*

Arundinaria amabalis: This species is popularly called A. amabalis and is native to the southwestern province of Guangdong in China. It also has an English name ‘token cane’, and it has been used in the production of flying fish rods in Europe and the Americas.

Table 1.6 Characterisations of some Genus

Genus/Taxon	Characteristics
Bambusa	***Brief Description***: This type of bamboo species are mostly medium large in dimension This genus is usually made up of about 157 different species which grows very well in the tropical as well as the subtropical zones across the globe with its culms wall being thick and rigid by nature. They grow from sea level up to 600–1200 m in altitude. A few species grow well in colder habitats with altitudes of 700–2000 m in a cold climate, while others grow well in the plains ***Geograpical Distribion***: Australia, Bangladesh, Bhutan, Burma, Pakistan, China, India, Japan, Kampuchea, Laos, Madagascar, Sri Lanka, Nepal, Papua New Guinea, the Philippines, Singapore, Thailand, Indonesia, Vietnam, Malaysia, and many others ***Soil type***: This type of bamboo grows in a wide variety of soil types but does grow well in a heavy textured well-drained sandy to clay loam soils. A well-drained residual soil also supports healthy growth. Other species grow well in alluvial soils and slightly acidic soils ***Areas of application***: They are used for construction purposes, pulp and paper, fuel, ornamental, instruments, handicraft, boards, composites and laminates, furniture, and other applications. Its shoots are edible
Borinda	***Brief Description***: This type of species of the genus bamboo is the largest of the temperate clump-forming Sino-Himalayan bamboos. They produce large, robust and durable culms with long internodes. They consist of about 11 different species. They grow from sea level up to 600 m in altitude ***Geographical Distribution***: These types of bamboo species are natives of Bhutan, Myanmar, China, Nepal, Tibet, Vietnam ***Soil type***: This kind of bamboo does well in a variety of soils, with loamy and moist soil being the most recommended. A well-drained residual soil also supports healthy growth. Other species grow well in alluvial soils and slightly acidic soils ***Areas of use***: They are used as fuel and ornamental plants and have many other purposes
Cephalostachyum	***Brief Description***: This genus can originally be found in the tropical and the subtropical zones of the globe with its culm wall being thin and weak in nature for applications as load-bearing elements. There are about 17 different species available in the tropical and subtropical regions where it grows well in moist soils, mostly in forests. This genus of bamboo can grow to heights ranging from 500 to 2000 m, with the Himalayas species reaching 2500 m and above ***Geograpical Distribution***: The native countries include Bangladesh, Bhutan, Mynamar, China, India, Madagascar, Nepal, the Philippines, Thailand, and Vietnam ***Areas of use***: They are used as fuel, ornamental plants, and the weaving of mats
Chimonobambusa	***Brief Description***: This genus is made up of 38 different bamboo species. The class of species of this genus are mainly running bamboo species characterised by swollen nodes which grow thorns ***Geograpical Distribution***: native to China, Japan, Vietnam, Myanmar, and regions of countries within the Himalayas ***Areas of use***: They are used as ornamental plants, walking sticks, food for pandas, and handicrafts or artefacts

(continued)

Table 1.6 (continued)

Genus/Taxon	Characteristics
Chusquea	***Brief Description***: This genus of bamboo is made up of 135 different species and are mainly found in the tropical and subtropical zones or regions of Central and South America. The culm of these species is characterised by being solid in nature, as compared to most bamboo culms which are hollow ***Geographical Distribution***: native to Chile, Mexico, and Argentina ***Areas of use***: They are used for musical instruments, walls, and furniture
Dendrocalamus	***Brief Description***: It is a subtropical bamboo species of this genus and are naturally clumpy, with thick culm walls as well as branches that are bulky in size. There are about 52 different species available in the subtropical regions. They grow well in colder habitats at altitudesof 700–2000 m in a cold climate, while others grow well on the plains. A few also grow from sea level up to 1200 m ***Geographical Distribution***: These types of bamboo species are natives of Bangladesh, Bhutan, Mynamar, China, India, Indonesia, Japan, Kampuchea, Laos, Malaysia, Nepal, Papua New Guinea, the Philippines, Sri Lanka, Thailand, and Vietnam ***Areas of use***: This species of bamboo is used for heavy- duty construction, building structures, floor coverings, reinforcement of concrete, scaffolding, and tiles. Its shoots are edible
Fargesia	***Brief Description***: The bamboos in this genus are made up of about 84 known species, which are clumpy in nature. They are among the world's most hardy bamboos, with thick clumping habits. Some of its species are food for pandas ***Geograhpical Distribution***: Mynanar native to Mynamar, China, and Vietnam ***Areas of use***: It is used as plant supports and woven into mats and has many other uses
Gigantochloa	***Brief Description***: This is a tropical genus of giant clumping species, similar to the genus Bambusa. It comprises about 55 different species recognised in the tropical regions ***Geographical Distribution***: grow well in South and Southeast Asia, including Bangladesh, Mynamar, Brunei, China, India, Indonesia, Laos, Malaysia, the Philippines, Singapore, Thailand, and Vietnam ***Areas of use***: It is used for handicraft, furniture, crafts, fuel, musical instruments, and for other purposes
Guadua	***Brief Description***: The species of this bamboo genus contain the largest bamboos in tropical America. Its stem can reach up to 30 m in height and 20 cm in diameter. It consists of about 38 species which grow well in the tropical regions. These species of bamboo are usually found in hot lowland regions, within forests and savannas ***Geograpical Distribution***: grow well in Central and South America, including Argentina, Mexico, and Paraguay ***Areas of use***: It is used for building and construction works, boards, artefacts or handicrafts, pulp, and household utensils, plus many other applications

(continued)

Table 1.6 (continued)

Genus/Taxon	Characteristics
Himalayacalamus	***Brief Description***: There are about 11 species available in this genus of bamboo, which are described as clumpy bamboos. The grow at low altitude in the Himalayas. This genus species looks familiar to Drepanostachyum in nature, but the distinction is in the branches. Drepanostachyum has many branches, while Himalayacalamus species have only one dominant branch ***Geograpical Distribution***: grow well in India, Nepal, Bhutan, and China ***Areas of use***: Some are used for furniture, scaffolding, and handicrafts, and for many other purposes
Indocalamus	***Brief Description***: There are about 35 distinct species of this bamboo genus ***Geographical Distribution***: native to China, Japan, Vietnam, and Sri Lanka ***Areas of use***: Its culm is used for containers in gardens, artefacts, handicrafts, for land rehabilitation, riverbank stabilisation, checking of erosion, plus many other applications
Nastus	***Brief Description***: This is a genus of slender, erect, scrambling, or climbing bamboo. This species of bamboo is mainly found in the tropical mountain forests that range throughout the Southern Hemisphere. It comprises 24 species ***Geographical Distribution***: native to Indonesia, Madagascar, Papua New Guinea, and the Solomon Islands ***Areas of use***: It is used for construction, furniture, crafts, scaffolding, paper, pulp, and for other applications. Its shoots are edible and are of high quality
Otatea	***Brief Description***: This species of bamboo is a genus of clumping bamboos. It comprises two different species and grows well on gentle slopes. It grows at altitudes between 200 and 2000 m ***Geograpical Distribution***: native to Central America, including Costa Rica, El Salvador, Honduras, and Nicaragua, and Mexico ***Areas of use***: Some of its uses include landscaping, land rehabilitation, ornamental plants, fencing, furniture and joinery, handicrafts, and for many other purposes
Oxytenanthera	***Brief Description***: This species of bamboo is a genus of clumping bamboos that is commonly available in most of sub-Saharan Africa. It comprises two different species, grows well in well-drained soil, and can be resist drought. It grows at an altitudes between 1200 and 1800 m ***Geographical Distribution***: native to Africa, including Ethiopia, Malawi, Ghana, Zambia, and Zimbabwe ***Soil type***: This type of bamboo can survive or do well in different kinds of soils ranging from loamy to moist soils. A well-drained residual soil also supports healthy growth ***Areas of use***: Some are used for the construction of structures, furniture, handicrafts, bamboo composite boards, scaffolding paper and pulp, and for many other purposes
Pseudosasa	***Brief Description***: These species of bamboo are small to medium running bamboos, which comprise about 36 different species of bamboo ***Geographical Distribution***: native to China, Japan, and Korea ***Areas of use***: They are used for fencing, walling, containers, and weapons such as arrows, and many other purposes

(continued)

Table 1.6 (continued)

Genus/Taxon	Characteristics
Schizostachyum	***Brief Description***: These bamboo species of the genus are native to tropical and subtropical regions and can be described as clump forming types. About 45 different species of this genus have been introduced into the horticulture industry for decoration This Schizostachyum species of bamboos usually grow spontaneously up to altitudes of 1000 m, with others growing up to altitudes of 2000 m in moist soils as well as in the forest zones. There are others that can grow up to about 2000 m, in forests and also in open fields where they prefer humid conditions ***Geographical Distribution***: native to Madagascar and Southeast Asia (the centre of distribution) ***Areas of use***: Their uses includes woven baskets and mats, musical instruments, small containers, and fishing rods, and for many other purposes
Semiarundinaria	***Brief Description***: These genera of bamboo can be located in the subtropical and temperate regions. They are classified as running bamboos and are made up of seven different species ***Geographical Distribution***: native to China and Japan ***Areas of use***: The use of its culm includes supporting plants, as ornamental plants in gardens, riverbank stabilisation, and for many other applications. Also, its shoots are edible
Thamnocalamus	***Brief Description***: This species of the bamboo genus is described as a clumping bamboo, which consists of about four different species. These genera are similar to Fargesia and are sometimes identified in synonymy with Thamnocalamus ***Geographical Distribution***: native to Bhutan, China, India, Madagascar, and Nepal, and to Southern Africa countries ***Areas of use***: The use of its culm includes construction of low-value artefacts, handicrafts, roofing, walling, woven mats, riverbank stabilisation, land rehabilitation, and walls, and has many other uses
Yushania	***Brief Description***: These species of bamboo genus range from temperate to subtropical regions. Its culms' height ranges between 2 and 20 m, with a diameter range of 5–13 cm. This species has a wall thickness of 1 cm. It consists of about 87 different species of bamboo and can grow at an altitude of 3000 m. These species are characterised by features such as long-necked rhizomes. These species of bamboo are semi-running and spread even further when growing in the right environment ***Geographical Distribution***: native to Malawi, Burundi, Ethiopia, Kenya, Tanzania, DR Congo, Madagascar, Cameroon, the Philippines, and The Republic of China ***Areas of use***: The use of its culm includes construction of low-value artefacts, handicrafts, rehabilitation of degraded lands, checking of erosion along the riverbank, and walls of buildings, and for many other applications

Adopted from Guadua Bamboo Company (20-11-2016)

Table 1.7 Some important bamboo species and their characteristics

Species	Characteristics
Bambusa arundinacea (*B. bambos*)	***Brief Description***: This bamboo species is a very tough, vigorous, and widespread, versatile bamboo. It has a thick culm internodes making it tough to split, and this restricts its value when being used to produce products of high quality. This species is thorny in nature with its culms rising up to 30 m, with an 18-cm diameter, wall thickness of 2.5–5 cm, and internodes up to 40 cm. It is native to tropical and subtropical regions. This species of bamboo usually grows up to about altitudes of 1200 m and can withstand a temperature of 2 °C. Its clumps are variable, with some culms being straight in nature while other are thornless. The lower branches of this species bear recurved spines, which are dense, thicker, and impenetrable. The clumps of this species flower, and this could be either the whole clump or a portion of the culms. The flowering of this species produces large amounts of seeds, indicating high levels of diversity for propagation in the next generation ***Geographical Distribution***: native to Bangladesh, China, India, Myanmar, and Thailand and introduced into other nations such as Vietnam, the Philippines, Indonesia, Nepal, and many others. ***Areas of use***: The use of its culm includes construction of low-value artefacts, for manufacturing of pulp and paper, handicrafts, for land rehabilitation, riverbank stabilization, and for many other purposes.
Bambusa balcooa	***Brief Description***: These are tropical and subtropical species that grow well in those zones. The culms are up to about 24 m tall with a diameter of 15 cm. The diameter of the nodes is about 30 to 45 cm with a wall thickness of 2.5 cm. This species of bamboo also flowers to produce seeds occasionally, and these can be used to propagate for the next generation, as well as its culms and offsets ***Geographical Distribution***: native to Bangladesh and India and introduced into other nations such as Australia and Indonesia and many others ***Areas of use***: The use of its culm includes construction of low-value artefacts, for manufacturing of pulp and paper, handicrafts, for furniture, and for many other purposes
Bambusa blumeana	***Brief Description***: It is a giant thorny tropical and subtropical bamboo which has limited commercial value. It is densely tufted, with its culms up to about 15 to 25 m tall with a diameter of 20 cm. The internodes range from 25–60 cm in length. It is green and glabrous with prominent nodes. This species of bamboo also produces gregarious flower within 20–30 years. However, it also flowers sporadically. They are propagated by using seeds, culm cuttings, layering, marcotting, and rhizome cuttings ***Geographical Distribution***: native to Borneo, Java, and Sumatra, and introduced to nations such as China, Thailand, the Philippines, Malaysia, and Papua Guinea

(continued)

Table 1.7 (continued)

Species	Characteristics
	Areas of use: The use of its culms includes construction of low-value artefacts and furniture, for manufacturing of pulp and paper, handicrafts, chopsticks, for rehabilitation of degraded land, riverbank stabilization, as well as other uses
Bambusa polymorpha	***Brief Description***: The culms' height ranges between 16 and 25 m with a diameter of 15 cm, and it grows well in the temperate zones. This species has a wall thickness of 1 cm. This species of bamboo also flowers to produce seeds periodically, between 50 to 60 years. It is propagated using seeds, rhizome cuttings, marcotting, layering, culms, and offset cuttings ***Geographical Distribution***: native to Bangladesh, India, Myanmar, and Thailand ***Areas of use***: The use of its culms includes construction of buildings, artefacts or handicrafts, and for furniture. Also, its shoots are edible
Bambusa textiles	***Brief Description***: These species of bamboo are medium sized and found in temperate and subtropical regions, with a 15-m culm height. The culm diameter is between 3 cm to 5 cm, while its internodes range from 60 to 80 cm long having a thin, delicate wall. Although a rarity, this species of bamboo also flowers to produce seeds. It is propagated using seeds, culms, and offset cuttings ***Geogrgraphical Distribution***: native only to China ***Areas of use***: The use of its culm includes a high-quality bamboo for weaving purposes since its culms split easily, and also for artefacts and handicrafts. Also, its shoots are edible
Bambusa tulda	***Brief Description***: These are temperate and subtropical bamboo species. Its culm diameter ranges 5–10 cm with a height of 30 m. This species has a culm wall thickness of 1 cm with an internode length ranging 40–70 cm. This species of bamboo also flowers to produce seeds periodically, between 25 to 40 years. It is propagated using seeds, marcotting, marcroprolififeration, rhizome cuttings, culms, and offset cuttings ***Geographical Distribution***: native to Bangladesh, India, Myanmar, and Thailand, and introduced into other countries such as Indonesia, Thailand, Nepal, and Vietnam, and many others ***Area of use***: The use of its culms includes artefacts of handicrafts, for construction and architectural works, furniture, and for paper and pulp
Bambusa vulgaris	***Brief Description***: These species of bamboo are strong and medium to large, with relatively open clumps and are found in temperate, tropical and subtropical regions. They have culms which are not straight, and the internodes are often curved. These species are of two kinds, namely B. vulgaris 'vittata' and B. vulgaris 'wamin', the former having yellow culms and the latter having green culms. They are grown as ornamental plants.

(continued)

Table 1.7 (continued)

Species	Characteristics
	The culm diameter is between 5 and 10 cm with a height of 20 m This species has a culm-wall thickness of 1.5 cm with an internode length ranging from 25 to 35 cm. This species of bamboo possesses the characteristic of having sporadic and gregarious flowering. They are propagated mostly using branch cuttings, marcotting, marcroprolififeration, rhizome cuttings, culms, offset cuttings, and layering. These species of bamboo are tremendously vegetative and easy to propagate ***Geographical Distribution***typical of all regions where bamboo is grown ***Areas of use***: The use of its culms includes artefacts or handicrafts, for construction and architectural works, and for furniture, pulp, and paper
Cephalostachyum pergracile	***Brief Description***: These species of bamboo are medium sized, with straight culms that keeps the culm sheath, and they are found in tropical and subtropical regions. Its culm height is up to 30 m, with a thin culm wall and an internode 45-cm long. This species of bamboo does not gregariously flower. They are propagated mostly using culm cuttings, offset cuttings, and seeds. These species of bamboo are tremendously vegetative and easy to propagate ***Geographical Distribution***: native to China, India, Myanmar, Java, Thailand ***Areas of use***: The use of its culms includes handicrafts, as an ornamental plant, for the construction and weaving of baskets, and for other artefacts
Dendrocalamus asper	***Brief Description***: These are subtropical and tropical bamboos which possess large culms with their diameters ranging between 8 cm to 20 cm, as well as heights ranging 20–30 m. Its culms' height range 20–30 m, with a diameter ranging from 8 to 20 cm. This species has a culm wall thickness of 2 cm with an internode length ranging from 20 to 45 cm. The gregarious flowering of this species of bamboo is not defined. It is propagated using culm cuttings, offset cuttings, and branch cuttings ***Geographical Distribution***: native to Bangladesh, India, Laos, Myanmar, Nepal, Thailand, and Vietnam and introduced into China, the Philippines, Indonesia, and Malaysia, as well as many other countries ***Areas of use***: The use of its culms includes artefacts or handicrafts, for construction and architectural works, furniture, and musical instruments, and for many other purposes. Also, its shoots are edible
Dendrocalamus giganteus	***Brief Description***: These are tropical and subtropical bamboo species and are made up of large and huge culms. Its culm height ranges from 25 to 60 m tall, with a diameter ranging from 10 to 20 cm. It has green to dark bluish-green features. This species has a culm wall thickness of 2.5 cm, with an

(continued)

Table 1.7 (continued)

Species	Characteristics
	internode length ranging from 40 to 50 cm. This species is not known for gregarious flowering. It is propagated using culm cuttings, rhizome cuttings, branch cuttings, macroproliferation, marcotting, and layering ***Geographical Distribution***: native to Myanmar and Thailand but have been introduced into other nations such as China, India, Bangladesh, Ghana, Thailand, Vietnam, Indonesia, Kenyan, and the Philippines, as well as many others ***Areas of use***: The use of its culms includes for building, structural works, bamboo boards, artefacts or handicrafts, for pulp, and for many other purposes. Also, its shoots are edible
Dendrocalamus lactiferous	***Brief Description***: These are tropical and subtropical zone bamboos and have mainly medium-sized culms. They grow very well in areas with high rainfall. The culm height ranges from 14 to 25 m, with a diameter ranging from 8 to 20 cm. This species has a culm-wall thickness of 0.5 to 3 cm and an internode length ranging from 20 to 70 cm. The gregarious flowering of this species of bamboo is not known. It is propagated using culm cuttings, layering, and marcotting ***Geographical Distribution***: native to China, Myanmar, and Taiwan. However, and introduced into countries such as India, Japan, Vietnam, the Philippines, and Thailand, as well as many others ***Areas of use***: The use of its culm include; artefacts or handicrafts, for construction and architectural works, furniture, and pulp, and for many other purposes
Dendrocalamus strictus	***Brief Description***: They are tropical and subtropical bamboo species which have medium-sized culms. The culm height ranges from 8 to 20 m with a diameter ranging from 2.5 to 8 cm. This species has a thick wall but is not straight and with an internode length ranging from 30 to 45 cm. This species is known to have a flowering cycle ranging between 20cand 40 years. It is propagated using seeds, culm cuttings, rhizome cuttings, macroproliferation, marcotting, and layering ***Geographical Distribution***: native to India, Myanmar, Nepal, and Thailand, and some other Asian countries ***Areas of use***: The use of its culms includes for building, structural works, bamboo boards, artefacts or handicrafts, for pulp, and for many household utensils. Also, its shoots are edible
Gigantochloa apus	***Brief Description***: These species of bamboo are large and are found in tropical regions. The culm height ranges from 8 to 30 m with a culm diameter ranging from 4 to 13 cm. This species is strongly tufted with a wall thickness of 1.5 cm and an internode length ranging from 35 to 45 cm. There is no known report of having a gregarious flowering cycle. It can be propagated by culm cuttings and offset cuttings

(continued)

Table 1.7 (continued)

Species	Characteristics
	Geographical Distribution: native to Malaysia, Myanmar, Thailand, and Indonesia, introduced to India and many other countries ***Areas of use***: The use of its culm includes for building, structural works, artefacts or handicrafts, for pulp, and for many household utensils and musical instruments. Also, its shoots are edible. However, it has a very bitter taste
Gigantochloa levis	***Brief Description***: These species of tropical bamboos have large culms and are found in tropical and subtropical regions. Its culm height is 30 m tall, with a diameter ranging from 5 cm to 16 cm. This species has a culm-wall thickness being 1–1.2 cm and an internode length of 45 cm long. The gregarious flowering of this species of bamboo is not known. It can be propagated using culms cutting and offset cutting ***Geographical Distribution*** native to the Philippines and Indonesia, including Kalimantan, as well as Malaysia, China, and Vietnam ***Areas of use***: The use of its culms includes artefacts or handicrafts, for construction and architectural works, furniture, walls, for household utensils, pulp, and paper. The shoots of this species have a highly edible quality
Gigantochloa pseudoarundinacea	***Brief Description***: This species are found in the tropical and subtropical zones. Its culm height ranges between 7 and 30 m, also having a diameter ranging between 5 and 13 cm. This species has a medium-to-thick, strong culm-wall of 2 cm, with an internode length ranging from 35 to 45 cm. This species is not known for a gregarious flowering cycle. It can be propagated using culm cuttings and branch cuttings ***Gegrapical Distribution***: naïve to Java, Sumatra, and Malaysia; however, it has been introduced into China, India, and other countries ***Areas of use***: The use of its culms includes structural works, artefacts or handicrafts, toothpicks, and for many household utensils. Also, its shoots are edible
Guadua angustifolia	***Brief Description***: These species of bamboo are large, are distributed in tropical regions, and can easily be identified by their aesthetic features such as dark green culms with white node bands. The diameter of its culms is about 20 cm, with a height of 30 m. This species is strongly tufted with a wall thickness of 1.5 cm and an internode length ranging from 35 to 45 cm. There are no records of having a gregarious flowering cycle. It can be propagated by culm cuttings and offset cuttings ***Gegraphical Distribution*** native to Argentina and Mexico, but have been introduced to China and India, as well as many other countries ***Areas of use***: The use of its culm includes for building, structural works, artefacts or handicrafts, for pulp, furniture, construction, and bamboo wood and boards

(continued)

Table 1.7 (continued)

Species	Characteristics
Melocanna baccifera (*Grove Forming*)	***Brief Description***: These species of tropical bamboos have large culms and are ecologically distributed across the tropical and the subtropical. The culm height ranges from 10 to 25 m, with a diameter ranging from 5 to 9 cm. This species has a thin culm wall of 0.5–1.2 cm and an internode length ranging from 20 to 60 cm. The culm tips are pendulous and produce the largest fruit in the grass family. The gregarious flowering of this species of bamboo takes between 45 and 50 years. It can be propagated using culm cuttings and seeds ***Geographical Distribution***: native to Bangladesh, India, and Myanmar and also now being grown in China, Indonesia, and Vietnam, as well as many other countries ***Areas of use***: The use of its culm includes handicrafts, for the construction of roofings, the weaving of mats, structural works, walls, pulp, and paper. Also, its shoots are edible and are also used for the preparation of liquor
Phyllostachyspubescens (*Grove Forming*)	***Brief Description***: These species of bamboo are medium to large and are found in tropical and subtropical regions. The culm height ranges from 10 to 20 m, with a diameter ranging from 18 to 20 cm having a waxy white covering. This species has an internode length of up to 45 cm. This species is not known for a gregarious flowering cycle, but it does flower sporadically. It is propagated using offset cuttings and seeds ***Geographical Distribution***: native to China. and introduced into Europe, Korea, Vietnam, Japan, USA, and other countries ***Areas of use***: The use of its culm includes construction, structural works, handicrafts, and for many other purposes
*Ochlandra*spp.	***Brief Description***: These species of bamboo are extensive and are found in tropical regions. The culm height ranges between 5 and 10 m, with a diameter of 5 cm. Flowering has been recorded every seven years of the life cycle. It can be propagated by culm cuttings ***Geographical Distribution*** indigenous to India and Sri Lanka, as well as many other countries ***Areas of use***: The use of its culms includes artefacts or handicrafts, walling, pulp and paper, and for many other applications
Thyrsostachys siamensis	***Brief Description***: These species of bamboo are densely clumped and are found in tropical and subtropical regions. The shealth on the culm of this species are obstinate, with a white ring below its nodes, and also has heights ranging between 8 m to 16 m tall, with a diameter ranging from 2 to 6 cm. This species has a thin culm wall with an internode length ranging between 15 cm and 30 cm. The species is adaptable to wet areas in healthy soils. This species of bamboo produces both sporadic and gregarious flowering within 48 years. It can be propagated using offset cuttings, seeds, and macroproliferation

(continued)

Table 1.7 (continued)

Species	Characteristics
	Geographical Distribution: native to Indochina and Myanmar, and introduced into many others countries for propagation ***Areas of uses***: The use of its culms includes handicrafts, construction, walls, structural works, pulp and paper, furniture, and as ornamental plants. Also, its shoots are edible

Adapted from www.guaduabamboo.com Bamboo Company (20-11-2016)

Bambusa atra: This species of bamboo, known as B. Atra, is mainly found in eastern Indonesia where it can grow to a height of 10 m.

Bambusa heterostachya: This species of bamboo grows to a height of 16 m, with a thick culm wall. The length of its internodes ranges from 30 cm to 80 cm. These species are distributed in countries such as Indonesia and Malaysia.

Bambusa nutans: The culms of this species have internodes of 45-cm lengths and grow to a height of 15 m. B. nutans bamboo species are natives of Thailand; application make it essential in the pulp industries as a raw material for pulp production, as well as the wood industry for the manufacturing of tables, chairs, and other furniture.

Bambusa chungii: This species of bamboo is described as a clumpy open type of bamboo, which originated in China; it is mostly applied in the horticulture field for decorations and as an ornamental plant. They have waxy, white aesthetic features on the culms.

Bambusa oldhamii: This form of bamboo species has its origin in China. It is a medium-sized bamboo which grows to a height of 12 m. It is mostly utilised for construction of structures, such as in furniture and many other architectural designs.

Bambusa pervariabilis: This species of bamboo is a medium-sized bamboo mostly utilised for agricultural tools and in the weaving and wood industry for basketry, mats, chairs, tables, and other furniture designs, as well as architectural works. It was originally found in Chinese provinces and in some other Asian countries.

Chusquea spp.: This bamboo genus has its origin in Latin America, and most of its culms are solid in nature. It consists of 200 species and finds application in diverse engineering fields.

Dendrocalamus brandisii: These species are native to India, China, and Myanmar. They are giant in nature, with a culm height of 35 m, with their diameter and the length of internodes being 20 and 40 cm, respectively. The fresh shoots of this species are highly edible. This species has found application in wood furniture, handicrafts, woven mats, and the basket industry.

Dendrocalamus hamiltonii: These species of bamboo are native to Asia with the greatest concentrations in China, India, Thailand, and Vietnam; the culms have a diameter of 18 cm, an internode length of 50 cm, and a height of 20 m. Their applications range from basket and mat weaving, handicrafts, firewood, and they also have highly edible shoots.

Dendrocalamus hookeri: This bamboo species originally grew in India, Myanmar, and Nepal. The culm of this species grows to the height of 20 m, with its diameter being 15 cm. It is mostly used for general construction and weaving purposes.

Dendrocalamus membranaceous: This bamboo species is native to Asia with China, India, Thailand, and Vietnam having the largest numbers. It has a culm height of 25 m, but it is thin in size, with a diameter of 10 cm. The plant is utilised in the construction, as well as the pulp and furniture, industries.

Gigantochloa albociliata: This species of bamboo mostly is native to China, India, Thailand, and other Asian countries where they grow to a height of 10 m. The fresh shoots of this species are highly edible, as well as the culms being used for furniture and construction works or designs.

Gigantochloa atroviolacea: This species of bamboo originates from Java with its culm height being 15 m. They are mostly being utilised for handicraft works, as good edible fresh shoots, and for furniture construction.

Gigantochloa balui: The species are native to Indonesia, Thailand, and China where they reach a height of 12 m, with an internode length of 40 cm.

Gigantochloa hassarkliana: This species of bamboo has their origin from Indonesia and Thailand and has a fast growth rate. The culm diameter is 6 cm, and the culm height is 10 m. Its applications cut across the furniture industry, as well as the design and construction industry.

Guadua amplexifolia and Guadua chacoensis: This bamboo species is originally a native of Latin America, where it grows to a height of 20 m and has diverse applications ranging from architectural designs to construction.

Oxytenanthera abyssinica: This species of bamboo are mostly described as African bamboo with a thick-walled culm. It mostly grows in lowlands and can also withstand the drier seasons in most parts of the continent. Its applications cuts across musical equipment, weaving, and construction.

Phyllostachys spp.: These species of bamboo numbers over 100 with its utilisation mostly as ornamental plants and also for furniture designs. They predominantly originate from of China and Japan and can grow in any part of the globe, especially in the tropics and subtropics. The shoots of this bamboo are mostly highly edible. They are processed and canned in industry, while others are consumed fresh in China and Japan.

Schizostachyum spp.: These species of bamboo vary by sizes, ranging from small to medium size and have their origin from Asia, especially from Southeastern Asia. Its culms height is about 15 m, and most of this species are used for basketry weaving, mat weaving, furniture works, and many other constructional works.

References

Ahmed M, Kamke FA (2005) Analysis of Calcutta bamboo for structural composite materials: physical and mechanical properties. Wood Sci Technol 39:448–459

Alam MK, Sarker RH, Hassan MA (1997) Chemotaxonomic studies in peroxidase isozyme of bamboos from Bangladesh. Bangladesh J Bot 26(2):99–105

Amada S, Lchikawa Y, Munekata T, Nagase Y, Shimizu H (1997) Fiber texture and mechanical graded structure of bamboo. Compos B 288:13–20

Amada S, Untao S (2001) Fracture properties of bamboo. Compos B 32:451–459

Bamboo Phylogeny Group [BPG] (2012) An updated tribal and subtribal classification of the bamboos (Poaceae: Bambusoideae). In: Gielis J, Potters G (eds) Proceedings of the 9th world bamboo congress, Antwerp, Belgium, 10–12 Apr 2012, pp 3–27

Bouchenak-Khelladi Y, Salamin N, Savolainen V, Forest F, van der Bank M, Chase MW, Hodkinson TR (2008) Large multi-gene phylogenetic trees of the grasses (*Poaceae*): progress towards complete tribal and generic level sampling. Mol Phylogenet Evol 47:488–505

Bystriakova N, Kapos V, Lysenko I, Stapleton C (2003a) Distribution and conservation status of forest bamboo biodiversity in the Asia-Pacific region. Biodivers Conserv 12:1833–1841

Bystriakova N, Kapos V, Stapleton C, Lysenko I (2003b) Bamboo biodiversity: information for planning conservation and management in the Asia-Pacific region. In: UNEP-World Conservation Monitoring Center and International Network for Bamboo and Rattan, Cambridge, UK, pp 1–71

Chou CH, Hwang YH (1985) A biochemical aspect of phylogenetic study of Bambusaceae in Taiwan. III. The genera Arthrostylidium, Chimonobambusa and Dendrocalamus. Bot Bull Acad Sin 26(2):155–170

Clark LG, Zhang W-P, Wendel JF (1995) A phylogeny of the grass family (*Poaceae*) based on ndhF sequence data. Syst Bot 20:436–460

Clark LG (1997) Bamboos: the centerpiece of the grass family. In: Chapman GP (ed) The bamboos, Academic Press, London, pp 237–248

Clark LG, Dransfield S, Triplett JK, Sa′nchez-Ken JG (2007) Phylogenetic relationships among the one-flowered, determinate genera of Bambuseae (*Poaceae*: *Bambusoideae*). Aliso 23: 315–332

Clayton WD, Renvoize SA (1986) Genera graminum, Grasses of the world, vol XIII, Kew bulletin additional series. Her Majesty's Stationery Office, London, p 389

Dai L-M, Wang Y, Su D-K, Zhou L, Yu D-P, Lewis BJ, Qi L (2011) Major forest types and the evolution of sustainable forestry in China. Environ Manag 48:1066–1078

Das M, Bhattacharya S, Basak J, Pal A (2007) Phylogenetic relationships among the bamboo species as revealed by morphological characters and polymorphism analyses. Biol Plant 51:667–672

Dransfield S (1992) The bamboos of Sabah. Sabah forest records no. 14, Forestry Department, Sabah, Malaysia, p 94

Ferreira FM, Dorea MC, Leite KRB, Oliveira RP (2013) Eremitis afimbriata and E. magnifica (Poaceae, Bambusoideae, Olyreae): two remarkable new species from Brazil and a first record of blue iridescence in bamboo leaves. Phytotaxa 84:31–45

Fernandez EP, Avila Moraes M, Martinelli G (2012) New records and geographic distribution of Glaziophyton mirabile (*Poaceae*: *Bambusoideae*). Check List 8:1296–1298

Fisher A, Triplett JK, Ho CS, Schiller A, Oltrogge K, Schroder E, Kelchner SA, Clark LG (2009) Paraphyly in the Chusqueinae (*Poaceae*: *Bambusoideae*: *Bambuseae*). Syst Bot 34:673–683

Franklin DC, Bowman DMJS (2004) A multi-scale biogeographical analysis of *Bambusa arnhemica*, a bamboo from monsoonal northern Australia. J Biogeogr 31:1335–1353

Friar E, Kochert G (1991) Bamboogermplasmscreeningwithnuclearrestriction fragment length polymorphisms. Theor Appl Genet 82(6):697–703

Gadgil M, Prasad SN (1984) Ecological determinants of life history evolution of two Indian bamboo species. Biotropica 16:161–172

Goh WL, Chandran S, Lin RS, Xia NH, Wong KM (2010) Phylogenetic relationships among Southeast Asian climbing bamboos (*Poaceae*: *Bambusoideae*) and the Bambusa complex. Biochem Syst Ecol 38:764–773

Gopakumar B, Motwani B (2013) Adaptive strategies of reed bamboos, *Ochlandra* spp., to the Western Ghat habitats of India. Bamboo Sci Cult 26:33–40

Gratani L, Maria FC, Laura V, Guiseppe F, Digiulio E (2008) Growth pattern and photosynthesis activities of different bamboo species growing in a botanical garden of Rome. Flora 203:77–84

Grass Phylogeny Working Group [GPWG] (2001) Phylogeny and subfamilial classification of the grasses (*Poaceae*). Ann Missouri Bot Gard 88:373–457

Grass Phylogeny Working Group [GPWG] II (2012) New grass phylogeny resolves deep evolutionary relationships and discovers C4 origins. New Phytol 193:304–312

Guadua Bamboo https://www.guaduabamboo.com/uses/corrugated-bamboo-roofing-sheets. Accessed 20 July 2016

Guo ZH, Chen YY, Li DZ, Yang JB (2001) Genetic variation and evolution of the alpine bamboos (*Poaceae*: *Bambusoideae*) using DNA sequence data. J Plant Res 114:315–322

Guo ZH, Chen YY, Li DZ (2002) Phylogenetic studies on the Thamnocalamus group and its allies (*Gramineae*: *Bambusoideae*) based on ITS sequence data. Mol Phylogenet Evol 22(1):20–30

Guo ZH, Li DZ (2004) Phylogenetics of the Thamnocalamus group and its allies (*Gramineae*: *Bambusoideae*): inference from the sequences of GBSSI gene and ITS spacer. Mol Phylogenet Evol 30:1–12

Gyansah L, Kwofie S (2011) Investigation into the performance of bamboo using the notched and the un-notched specimen. Res J Appl Sci Eng Technol 3(4):245–251

Janssen JJA (1995) Building with bamboo, 2nd edn. Intermediate Technology Publication Limited, London, p 65

Janssen JJA (2000) Designing and building with bamboo. International Network for Bamboo and Rattan Technical Report No. 29. pp 207

Judziewicz EJ, Clark LG, Londoño X, Stern MJ (1999) American bamboos. Smithsonian Institution, Washington, DC, p 392

Judziewicz EJ, Clark LG (2007) Classification and biogeography of New World grasses: Anomochlooideae, Pharoideae, Ehrhartoideae and Bambusoideae. Aliso 23:303–314

Kelchner SA, Bamboo Phylogeny Group (2013) Higher level phylogenetic relationships within the bamboos (*Poaceae*: *Bambusoideae*) based on five plastid markers. Mol Phylogenet Evol 67:404–413

Larpkern P, Moe SR, Totland O (2010) Bamboo dominance reduces tree regeneration in a disturbed tropical forest. Oecologia 165:161–168

Latif AM, Liese W (2002) Culm characteristics of two bamboos in relation to age, height and site. In: Kumar A, Ramannuja IV, Rao AN, Sastry CH (eds), pp 223–233

Li DZ, Xue JR (1997) The biodiversity and conservation of bamboos in Yunnan, China. In: Chapman GP (ed) The bamboos. Academic, London, pp 83–94

Li DZ, Wang ZP, Zhu ZD, Xia NH, Jia LZ, Guo ZH, Yang GY, Stapleton C (2006) Bambuseae. In: Wu ZY, Raven PH, Hong DY (eds) Flora of China: *Poaceae*, vol 22. Science Press, Missouri Botanical Garden Press, Beijing, St. Louis, pp 7–180

Li H, Shen S (2011) The mechanical properties of Bamboo and vascular bundles. J Mater Res 26 (21):2749–2756

Lobovikov M, Paudel S, Piazza M, Ren H, Wu J (2007) World bamboo resources: an Athematic study prepared in the framework of the Global Forest Resources Assessment (2005) Rome-Italy. Food and Agriculture Organization (FAO) of the United Nations

Marchesini VA, Sala OE, Austin AT (2009) Ecological consequences of a massive flowering event of bamboo (*Chusquea culeou*) in a temperate forest of Patagonia, Argentina

Martins SV, Júnior RC, Rodrigues RR, Gandolfi S (2004) Colonization of gaps produced by death of bamboo clumps in a semideciduous mesophytic forest in southeastern Brazil. Plant Ecol 172:121–131

Nayak S, Rout GR, Das M (2003) Evaluation of the genetic variability in bamboo using RAPD markers. Plant Soil Environment 49:24–28

Nguyen HN (2006) Bamboos of Vietnam. Agricultural Publishing House

Ohrnberger D (1999) The bamboos of the world: annotated nomenclature and literature of the species and the higher and lower taxa. Elsevier, Amsterdam, p 585

Oteng-Amoako AA, Ofori DA, Anglaaere LCN, Obiri-Darko BE, Ebanyenle E (2004). Sustainable development of bamboo resources. Africa Forestry Research Network (AFORNET), Ghana Processing Report 1 (unpublished), Jan-Dec 2004, p 13

Oteng-Amoako AA, Ofori DA, Anglaare LCN, Obiri-Darko BE, Ebanyenle E (2005) Sustainable development of bamboo resources of Ghana and Togo. Progress report submitted to Africa forest research Network, Nairobi, Kenya

Chaowana P (2013) Bamboo: an alternative raw material for wood and wood-based composites. J Mater Sci Res 2(2):90 (15 March 2013)

Pattanaik S (2008) Distribution, diversity and relationships in Dendrocalamus hamiltonii Munro: an appraisal using GIS and AFLP markers. Ph.D. thesis, School of Agricultural and Forest Sciences, Bangor University, Bangor, p 176

Pattanaik S, Hall JB (2011) Molecular evidence for polyphyly in the woody bamboo genus Dendrocalamus (subtribe Bambusinae). Plant Syst Evol 291:59–67

Ramanayake SMSD, Meemaduma VN, Weerawardene TE (2007) Genetic diversity and relationships between nine species of bamboo in Sri Lanka, using Random Amplified Polymorphic DNA. Plant Syst Evol 269:55–61

Ruiz-Sanchez E, Sosa V, Mejía-Saules MT, Londoño X, Clark LG (2011) A taxonomic revision of Otatea (*Poaceae*: *Bambusoideae*: *Bambuseae*) including four new species. Syst Bot 36: 314–336

Seethalakshmi KK, Kumar MSM (1998) Bamboos of India: a compendium. Kerala Forest Research Institute, Peechi, India and International Network for Bamboo and Rattan, New Delhi, p 342

Sen T, Reddy HNJ (2011) Application of sisal, bamboo, coir and jute natural composites in structural upgradation. Int J Innov Manage Technol 2(3) (June)

Sharma B, Gatóo A, Bock M, Mulligan H, Ramage M (2014) Engineered bamboo: state of the art. Proc Inst Civil Eng. doi:10.1680/coma.14.00020

Soderstrom TR (1981) Some evolutionary trends in the *Bambusoideae* (*Poaceae*). Ann Missouri Bot Gard 68:15–47

Soderstrom TR, Calderon CE (1979) V. ecology and phytosociology of bamboo vegetation. In: Numata M (ed) Ecology of grasslands and bamboolands in the world. VEB Gustav Fischer Verlag, Jena, pp 223–236

Soderstrom TR, Ellis RP (1982) Taxonomic status of the endemic South African bamboo, Thamnocalamus tessellatus. Bothalia 14:53–67

Soderstrom TR, Ellis RP (1987) The position of bamboo genera and allies in a system of grass classification. In: Soderstrom TR, Hilu KW, Campbell CS, Barkworth ME (eds) Grass systematics and evolution. Smithsonian Institution, Washington, DC, pp 225–238

Soderstrom TR, Ellis RP (1988) The woody bamboos (*Poaceae*: *Bambuseae*) of Sri Lanka: a morphological-anatomical study. Smithson Contrib Bot 72:1–75

Soderstrom TR, Londoño X (1988) A morphological study of Alvimia (*Poaceae*: *Bambuseae*), a new Brazilian bamboo genus with fleshy fruits. Am J Bot 75:819–839

Soderstrom TR, Judziewicz EJ, Clark LG (1988) Distribution patterns of neotropical bamboos. In: Vanzolini PE, Heyer RE (eds) Proceedings of a workshop on neotropical distribution patterns, Academia Brasileira de Ciencias, Rio de Janeiro, 12–16 Jan 1987, pp 121–157

Stapleton CMA (1994a) The bamboos of Nepal and Bhutan Part I: Bambusa, Dendrocalamus, Melocanna, Cephalostachyum, Teinostachyum, and Pseudostachyum (*Gramineae*: *Poaceae*, *Bambusoideae*). Edinb J Bot 51:1–32

Stapleton CMA (1994b) The bamboos of Nepal and Bhutan Part II: Arundinaria, Thamnocalamus, Borinda and Yushania (*Gramineae*: *Poaceae*, *Bambusoideae*). Edinb J Bot 51:275–295

Stapleton CMA (1994c) The bamboos of Nepal and Bhutan Part III: Drepanostachyum, Himalayacalamus, Ampelocalamus, Neomicrocalamus and Chimonobambusa. Edinb J Bot 51:301–330

Stapleton CMA (1998) Form and function in the bamboo rhizome. J Am Bamb Soc 12(1):21–29

Steinfeld C (2001) A bamboo future. Environmental Design and Construction, pp 1–5

Sun Y, Xia N, Lin R (2005) Phylogenetic analysis of Bambusa (*Poaceae*: *Bambusoideae*) based on internal transcribed spacer sequences of nuclear ribosomal DNA. Biochem Genet 43(11–12):512–603

Sungkaew S, Stapleton CMA, Salamin N, Hodkinson TR (2009) Non-monophyly of the woody bamboos (*Bambuseae*; *Poaceae*): a multi-gene region phylogenetic analysis of Bambusoideae s.s. J Plant Res 122:95–108

Taguchi-Shiobara F, Ishii T, Terachi T, Tsunewaki K (1998) Mitochondrial genome differentiation in the genus Phyllostachys. Jpn Agric Res Q 32(1):7–14

Triplett JK, Weakley A, Clark LG (2006) Hill cane (*Arundinaria appalachiana*), a new species of woody bamboo (*Poaceae*: *Bambusoideae*) from the Southern Appalachian Mountains. SIDA 22:79–95

Tsuyama I, Nakao K, Matsui T, Higa M, Horikawa M, Kominami Y, Tanaka N (2011) Climatic controls of a keystone understory species, Sasamorpha borealis, and an impact assessment of climate change in Japan. Annals of Forest Science 68 (4):689–699

Tyrrell CD, Santos-Gonc¸alves AP, Londoño X, Clark LG (2012) Molecular phylogeny of the arthrostylidioid bamboos (*Poaceae*: *Bambusoideae*: *Bambuseae*: *Arthrostylidiinae*) and new genus Didymogonyx. Mol Phylogenet Evol 65:136–148

Viana P, Filgueiras T, Clark LG (2013) A new genus of woody bamboo (*Poaceae*: *Bambusoideae*: *Bambuseae*) endemic to Brazil. Syst Bot 38:97–103

Wong KM (1991) The growth architecture and ecology of some tropical bamboos. J Am Bamboo Soc 8:31–46

Wong KM (1993) Four New Genera of Bamboos (Gramineae: Bambusoideae) from Malesia. Kew Bulletin 48 (3):517–532

Wong KM (1995) The bamboos of Peninsular Malaysia. Forest Research Institute Malaysia, Kuala Lumpur

Wong KM (2004) Bamboo, the amazing grass. A guide to the diversity and study of bamboos in Southeast Asia. International Plant Genetic Resources Institute & University of Malaya, Kuala Lumpur, p 80

Wooldridge M (2012) Booming bamboo: the next super-material? BBC News Magazine, 3 Apr 2012

Wu ZQ, Ge S (2012) The phylogeny of the BEP clade in grasses revisited: evidence from the whole-genome sequences of chloroplasts. Mol Phylogenet Evol 62:573–578

Yan Y, Benhua F, Bo Z, Xiang Y (2006) Cell-wall mechanical properties of bamboo investigated by in-situ imaging nanindentation. Wood Fiber Sci 39:527–535

Zhang W-P, Clark LG (2000) Phylogeny and classification of the Bambusoideae (*Poaceae*). In: Jacobs SWL, Everett J (eds) Grass systematics and evolution. CSIRO, Melbourne, pp 35–42

Zhang YX, Chun CX, Li DZ (2012) Complex evolution in Arundinarieae (*Poaceae*: *Bambusoideae*): incongruence between plastid and nuclear GBSSI gene phylogenies. Mol Phylogenet Evol 63:777–797

Zhou G, Meng C, Jiang P, Xu Q (2011) Review of carbon fixation in bamboo forests in China. Bot Rev. doi:10.1007/s12229-011-9082-z

Chapter 2
Regeneration, Cultivation, and Sustenance of Bamboo

Abstract This chapter examines the basic techniques available for regenerating seedlings to be used in plantations or on the farm. The three most important methods used in producing young seedlings for plantation are discussed in detail, as well as the processes for its cultivation. The chapter places emphasis on the regeneration of seedlings by seeds, culm cuttings, and rhizomes. The selection of land sites and the preparation for plantation of young seedlings is also looked at in this chapter, with a description of land selection, pre-planting operations, propagation methods, post-planting operations, harvesting, and handling. The effects of fungal infections and attacks by insects on the bamboo before and after harvesting are dealt with and also post-harvest treatment and preservation methods. The effects of these infections and attacks on the bamboo culms, when not treated or prevented in time, will result in the deterioration of the physical features of the culm. Various scientific forms of preservation and the types of preservatives used in its treatment are discussed. Sustainable techniques for bamboo plants are examined.

2.1 Introduction

Bamboo is a natural regenerative plant which naturally grows mostly in the forest as a bushy grass in tropical and subtropical ecology. It is also naturally found as an understory plant, which can also grow in moist regions and can be referred to as woody grass and classified into species ranging in number from 1439 to 1500 with 115 generas across the globe. The bamboo plant plays a significant role in the preservation of the forest as it releases about 35% more oxygen into the atmosphere than other plants, thus reducing atmospheric carbon dioxide as compared to hard and softwood. They are grown on farmlands in the plains and through the hills, as well as in the valleys. They also serve as a medium to check soil erosion because their remaining roots hold the surrounding soils after harvesting and provide nutrients to the topsoil. In recent years, the growth and uses of bamboo keep increasing rapidly across the globe due to its versatility, hence the need to encourage its commercial farming. The steady growth of its application over the

E.T. Akinlabi et al., *Bamboo*, DOI 10.1007/978-3-319-56808-9_2

past few years, in every aspect of humanity, has raised concern about how sustainable this plant will be over time since most industries are now competing for it as raw material for production and diverse applications globally.

Harvesting of bamboo from its native forests (naturally occurring), as well as those from commercial plantations, requires properly designed schemes for its harvesting, as this approach would enhance its sustainability. The farmers who go into commercial cultivation of bamboo need to be introduced to the appropriate techniques for propagation, as well as the harvesting techniques that promote the sustainability of bamboo plants. As new applications are being exploited industrially, scientific methods should be simply implemented to control its over-exploitation in farmlands, as well as the in natural forests. Over/exploitation is a global concern to industries and governments and needs discussion, despite bamboo's vigorous growth and somewhat invasive characteristics. There is a well-established and tested system of controlling this resource in countries such as China and India that can serve as a benchmark for its development. Many countries in Africa, such as Ghana and Ethiopia, have adopted the success story of these Asian countries by using the appropriate techniques for the development of bamboo and its products. Non-governmental organisations and the private sector are involved in its cultivation since it is in high demand for industrial purposes. Many farmers in Ghana have been encouraged to create commercial plantations of bamboo, as there is a high request for this resource for various industrial applications, especially in the design, manufacturing, and construction industries. Also, because bamboo can help to rehabilitate degraded lands, it is suggested that communities with degraded lands employ its propagation to restore the nourishment of the topsoils in communities of Ghana and other countries in Africa, as well as on a global scale.

Globally, commercial bamboo farmers are required to use an appropriate technology for the propagation of this resource in order to promote its sustainability. Bamboo does not depend strictly on inorganic chemicals such as insecticides, fertiliser and others for its growth in the natural forest. However, commercial farmers of bamboo do apply these chemicals for higher yields. Hence, appropriate methods of the application must be followed in order not to cause any negative effects on the land, as well as the entire ecosystem. Promoting bamboo conservation globally is recommended to make available (introduce) high-quality seedlings (breeds) or other planting stocks that can easily adapt to new environments (habitats) without much difficulty (Banik 1991).

2.2 Methods and Techniques of Bamboo Propagation

Bamboos can be propagated through the use of seeds (sexual) or by vegetative (asexual) methods, depending on the species available for plantations (Ramyarangsi 1988). However, not all bamboo species produce seeds, and even with those which do, the seeds are sporadic, as it takes about 30–120 years for some species to flower after which the parent plant dies off. Mostly in bamboo plants, the parents die off just after flowering, and this is associated with those species which flower gregariously. There are only a few bamboo species that flower frequently to produce seeds and can be used to propagate seedlings when required. However, bamboo seeds have a relatively short lifespan that prevents its preservation for longer periods, and, if a longer storage period is required, then an intense and sophisticated drying technique is adapted to enhance drying, after which they need to be sealed in airtight polybags or other vessels. Bamboo species that do not flower to produce seeds for regeneration of seedlings have to be propagated through vegetative techniques, such as culm (stem) cuttings or rhizomes of non-clump species (Rao et al. 1989; Fu and Banik 1996).

2.2.1 Bamboo Propagation

Bamboo can be propagated using basic planting techniques such as seedlings, vegetative propagation through culm cuttings, and rhizome transplants (Rao et al. 1989; Fu and Banik 1996; Banik 2008). However, before using any of these methods for plantations or farms, the following factors needs to be critically examined, among many others:

- The type of bamboo species to plant.
- Determination of the size of culm after maturity.
- Determination of the growth habit of the selected species.
- Planting style to ease movement for maintenance in the plantation.
- The harvesting technique.

The propagation of bamboo seedlings in a new site can be done by using either sexual (seedlings) or asexual techniques from the bamboo plants from nursery-raised seedlings, directly planting cut offsets of the culm, or by rhizome transplanting. These parts are nurtured to grow to form new seedlings. The growth rate of bamboo depends on the species type and the method used in the plantation, as well as the environment in which the plant is cultivated. The cultivation methods are classified as sexual and asexual propagation (Banik 1995). A well-drained soil is critical for bamboo growth and habitat, hence, before planting the seedlings, the land needs to be surveyed to know the nutrient level of the soil, as well as the porosity of the soil and its propensity to flooding. Bamboo has been observed not to do well in alkaline, as well as waterlogged, soils. Hence, recommendations are that, preferably, bamboo plantations/cultivation should be situated on moderate slopes (Rao et al. 1989; Fu and Banik 1996).

Plantations of bamboos with a high yield of culms are those with a well-planned layout for distances between seedlings. This plantation allows for free movements of persons and equipment within the plantation and for sunlight to reach every single plant on the farm (Rao et al. 1989; Fu and Banik 1996). It is important to plant seedlings during the rainy season and, when necessary, apply manure in the holes before planting seedlings to achieve high yields of bamboo culms. Bamboo seedlings need to be planted vertically in the holes, and the holes should be covered with the required amount of soil and then should be mulched. It has been observed that bamboo propagation highly depends on the species type, because some can only be propagated using culm cutting or rhizomes. In other words, some can also be propagated using seeds for high-quality yields of culms. Propagation of bamboo seedlings either by seed, culm, branch, or rhizomes cuttings requires a nursery, to enable regulation for a certain number of months or years, before being transplanted onto the farm.

2.2.2 Production of Bamboo Seedlings

The production of seedlings of bamboo is essential for regeneration of new propagules by using basic techniques, such as sexual (seeds) nursery, or asexual (culm cuttings and rhizome) transplants (Banik 1994). These production methods are discussed next.

2.2.3 Sexual Propagation

Sexual propagation is a reproduction method involved in producing new bamboo seedlings through the use of seeds (Rao et al. 1989; Fu and Banik 1996; Banik 2008). The seeds are necessary for propagation and reproduction of new seedlings. For large-scale plantations, bamboo seedlings are recommended as the suitable method of propagation because it has been scientifically proven that clumps produced from seeds tend to maintain their genetic originality (ancestral roots). Also, a high, maximum quality of culms is produced from a vegetation of clump bamboos which was originally propagated from seeds, and this can foster a longer life span for the clumps on the plantation. However, one characteristic of bamboos before seeds are produced is that the clumps die off after flowering. When seeds are produced, they have short-lived viability and hence, they should not be stored for long periods. To store them requires highly sophisticated drying, the correct temperature, and sealing techniques in a humidity environment. However, seeds stored under reduced humidity and temperature can only last for two years, after which they will not be viable for production of seedlings. Bamboo seeds have a high germination rate before deterioration if sown within s about three months. Bamboo seeds are first nursed in a nursery bed having a moist loamy soil. The seeds are sown by spreading them on the nursery bed after which the seeds are then covered with a required amount of moist soil to proper depth which can enhance its germination.

The nursery bed is required to be kept in shade, to protect the seedbed, as well as the newly germinating seedlings, from direct sunlight, and to provide a high rate of germination (Pattanaik et al. 2002). There should be constant watering to keep the seed beds, as well as the soil of the seedlings, moist. This activity should be performed in the early hours of the mornings or the evenings using fine rose-watering containers. To keep the soils moist on the nursery bed, but not have them in direct sun rays, the seeds in the nursery, as well as the seedlings, should be mulched in their initial months of nursing, thus protecting them. Mulching serves as a means of protecting the surface of the nursery bed from getting direct sun rays and from high impact by raindrops and rapid evaporation of soil moisture, as well as also to keep the soil under the seedlings moist. The primary materials usually used for mulching include; dry straws, grass, and leaves. The kind and the quantity of mulch depends on the time of sowing of the seeds in the nursery beds and the rate of growth of the seedlings. The nearly mature bamboo seedlings from the nursery can be transferred into Poly bags for full maturity before being transplanted at the farm or the plantation. See Fig. 2.1, a photo of a set of Poly bag pots ready to receive bamboo seedling for transfer from nursery beds in the process of sexually propagating bamboo.

The young bamboo seedlings are then uprooted from the nursery beds and transferred into polybags. The seedlings must be kept moist during this period of uprooting from the nursery as some amount of soil is attached to the seedlings. The process is done with much care so that the shoots of the seedling do not detach from the roots. A set of bamboo seedlings which are being transferred from the nursery bed to be planted into the row of polybags under the shade (Fig. 2.2).

Fig. 2.1 Rows of Poly bags ready to receive bamboo seedlings

Fig. 2.2 Bamboo seedlings being transferred from nursery into Poly bags

The stages of seedling growth labelled from A–D in Fig. 2.3 is explained here:

Stage A: Depicts the transplanting of the bamboo seedlings from the nursery into the row of Poly bags, which are kept under shade after one week

Stage B: Depicts the growth rate of the bamboo seedlings after one Month, under the shade in the row of Poly bags

Stage C: Depicts the growth rate of the bamboo seedlings after six weeks under the shade in the row of Poly bags

Stage D: Depicts the growth rate of the bamboo seedlings after eight weeks under the shade in the row of Poly bags.

Raising bamboo from seed is not as easy, as compared to rhizomes and stem cuttings. This is because most bamboos do not produce seeds, and those species that do flower and bear seeds take 30–100 years to bear flowers and subsequently seeds. The growing of bamboo from seed is, therefore, rare, and it is only through vegetative propagation methods that seedlings can be grown or propagated by growers (Banik 1994).

2.2.4 Asexual Propagation

This method of propagation makes use of the vegetative parts which includes rhizomes, offsets, and culm cuttings. These propagules are then carried to the nursery for regeneration of new shoots. The formation of new shoots from these propagules is allowed to mature and develop new rhizomes before being transferred

Fig. 2.3 Stages of growth of bamboo seedlings in Poly bags under shade

to their permanent place of cultivation. The types of vegetative propagation methods include offset planting, rhizome planting, culm cut planting, split culm cuttings, branch cuttings, layering, marcotting, and macro proliferation of seedlings (Banik 1994, 1995).

2.2.4.1 Bamboo Propagation by Offset Cuttings

The offset method of culm propagation requires the usage of a length of culm possessing between 3 and 5 nodes from the bottom (approx. 1–2.5 m) and having the rhizome axis and the roots attached. Offset propagation is said to be one of the appropriate methods of regeneration of bamboo seedlings. The culm employed in this technology should be between one to two years old (Rao et al. 1989).

Offset Cutting

The culm of bamboo is cut at a slant angle to a suitable length, after which it is uprooted from the ground with its roots and rhizomes intact, for the regeneration and development of new shoots (Rao et al. 1989). To ensure their survival in the field of the plantation, the following should be considered:

A healthy parent clump should be the source for collecting the offsets of bamboo and:

- One must be cautious not to damage the rhizomes and the attached roots during offset cutting.
- The rhizome is separated from the clump during offsetting, and usually the culm is taken from the outer ends of the clump.
- The buds on the rhizome are lightly pressed to see that they are not rotten.
- After uprooting the offset rhizome from the ground, it is important to cover the base portion of the offset culm in Poly bags or jute bags so as to have moist soil or wet sawdust filled in before being transported to the place of cultivation.

Treatments

The application of offsets culms of bamboo for cultivation on a plantation requires some basic treatments and technicalities before being transferred to the farm. It is advisable that, during the offsetting of culms from the parent clumps, care should be taken to make sure roots and rhizomes are not damaged in the process. Also, offsets materials need to be available at the planting site as early as possible before the rainy season starts. Planting of the offset should be done at the beginning of the rainy season to enhance the chances of survival. Before transplanting the offsets of the bamboo, it is important to collect them as early as two to three months before the planting season (Midmore 2009). The offsets can be collected and kept in a nursery bed close to the plantation temporally in a moist and loamy soil. The slant cut offset of culm is sealed with earth. When offsets are planted, it is mandatory to isolate the plants from other plants that may be competing for soil nutrients (Midmore 2009). Therefore, clearing of weeds and other plants from the plantations is necessary, using weedicides or using manual cutlasses and hoes to maintain the farm. The chance of survival of offsets plantation is 50% during an outbreak of drought, and, in such periods, watering is required for the plants' sustenance.

Limitations

- Offsets are bulky and weighty in a range of 6–30 kg.
- This method of planting is costly due to its labour intensity. Excavation and transportation to the site are expensive.
- One can only excavate a limited number of bamboo offsets from a clump. Harvesting of more than two to three offsets in some cases from a single clump may lead to regeneration of the parent clump and retarded growth until, at a certain time, it dies off completely.
- The method is unsuitable for developing large plantations.

- This method of propagation can only be employed in bamboo clumps forming species with thick walls (i.e. Melocanna baccifera).
- This method of propagation shows a 30–50% rate of success for Melo-canna baccifera and other bamboo species.

2.2.4.2 Bamboo Culm Cutting Propagation

Bamboo culm cuttings for propagation are segments of the culm (stem) usually consisting of one or up to between two–three nodes with buds or branches. This method of propagation applies to most clump-forming bamboo species and might not be suitable for species that are non-clumpy. Using this method of propagation is efficient when various pieces of bamboo culms are cut with a single or more nodes. The culm cuttings are then placed horizontally in the propagation beds in the nursery with the vascular culm tissue slanted into the wet soil at an angle (Rao et al. 1989; Zamora 1994). The nursery bed is covered with a shade cloth and is constantly watered until the cuttings sprout and the roots are well developed or mature up to a certain stage when they are then transferred into Poly bags. In this process, the nodes with more than one shoot and a well-established root system are carefully split into two or more divisions before being transferred into the Poly bags (Fig. 2.4). The bagged seedlings are then kept under the shade and are constantly watered until they are fully mature; they are then moved to the farm or another place of cultivation for transplanting. This form of propagation production can produce several shoots.

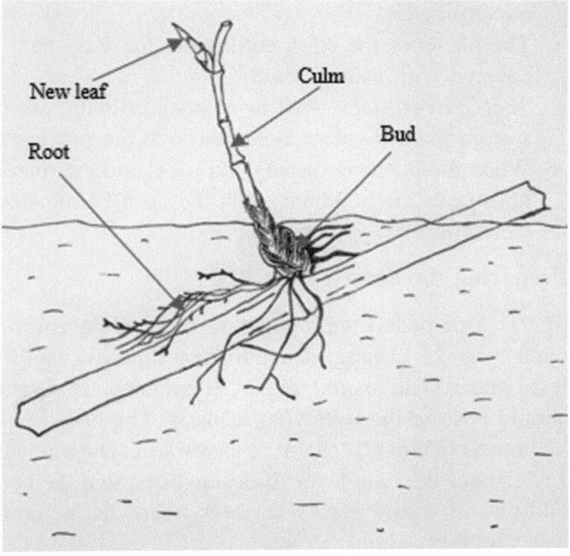

Fig. 2.4 Bamboo shoots propagation using culm cutting process

Bamboo Culm Cuttings

One of the methods used in germinating bamboo seedlings is from culm cuttings; the steps are as follows (Rao et al. 1989; Zamora 1994; Midmore 2009):

- A vigorous and healthy culm is selected for the cutting, preferably not more than two years of age.
- The parts employ for cutting are taken from between the base and the middle portion of the bamboo culm.
- The first cut of the culm should be about one–two inches below the node.
- The second cut should also be about one–two inches below the next node.
- The branches must be trimmed off, leaving only one or two at the node end.
- The part of the branches of the culm selected for propagation is cut to a length of 10–30 cm. Care must be taken not to damage the new bud with any sharp edge tool.
- The parts are cut with a sharp knife or saw. Care should be taken to prevent the culm from splitting during the cutting process, especially culms that have thin walls. There should be an allowance of 5–10 cm kept on both sides of the node on the culm cutting.
- The cut end of the culm is waxed immediately after cutting or segmentation and also covered with jute sackcloth which needs to be moistened or by being placed in loamy soil, thus reducing evaporation of water from the cut ends.
- The bamboo cuttings are then transported to the propagation or nursery bed
- The culm cuttings are then nursed in the bed or in Poly bags containing moist loamy soils.
- The bamboo culms are then inserted into the ground at an angle, covering the node and leaving the branches sticking out of the ground.
- A culm cut plants are watered constantly and regularly in the Poly bags or the nursery bed.
- The plants or the culm cuttings in the Poly bags or the nursery bed should be covered with a shade cloth.
- The culm cuttings need to be covered with soil occasionally when exposed to sustain the ground moisture through the process of mulching.
- When the bamboo sprouts and roots and the rhizomes are well developed in the nursery or the polybags, then they can be transferred to the farm or the site for cultivation.

Preparing the Propagation Bed

The size of beds for propagation of bamboo culm cuttings said to be one-meter wide by 5–15 m long, depending on land availability. Beds must be formed on a level ground and should be 20–30 cm deep. The ground or soil in the prepared bed should possess the following features. The bed should be covered with loamy soil and must contain 10–15 cm of sandy soil. The top layer of fine sand must range 10–15 cm over the base layer. It is suggested that the beds be three-layered to aid easy outflow of rainwater, whenever there is a continuous rainfall, to prevent water-logging (Banik 1994).

Placing Culm Cuttings in the Bed

The placement of bamboo culm cuttings in the prepared bed is required to be horizontal, and the space between culms must be in the range 15–30 cm. The process also requires that the culms be placed in the top soil layer, ranging from three–five cm above the sandy soil and then covered by three–six cm of the fine-sand soil finally. The bed is mulched to keep the soil moist, while the plants are kept under partial shade to have some degree of sunlight reaching them.

In the case where the internode is long (e.g. *Bambusa polymorpha* or *Schizostachyum lima*), the segment may have one–two nodes. Also, when the internode is short, segments with three nodes may be prepared for this propagation. For higher regeneration of seedlings, there must be more nodes on the culm for propagation. However, culm cuttings that are long create problems in handling and transportation.

2.2.4.3 Bamboo Propagation with Rhizomes

This process of propagation has been traditionally used to propagate non-clump forming bamboo species, where rhizomes are slightly leptomorph and long. These are rarely used to regenerate or propagate clump-forming species seedlings of bamboo with pachymorph rhizome types. The rhizome is an underground stem, which consists of two parts, namely the actual rhizome and the rhizome neck, which is the base of the original rhizome. The stem of the rhizome has nodes, but those of the neck always lack buds and usually roots (Zamora 1994). The original rhizome has roots or root primordia and buds at all of the nodes. There are two distinct types of rhizomes, which are called pachymorph and leptomorph.

Pachymorph Rhizomes: These rhizomes are characterised by a short, thick, curved and sub-fusiform and are rarely spherical in shape. The internodes of pachymorph rhizomes are broader and asymmetrical (i.e. larger on the side that bears buds). The shoots at each node are in a row, usually five each on the side of the rhizome. When pachymorph rhizomes are dormant, the bud is dome-shaped but asymmetrical (Rao 1989). The pachymorph rhizomes are distinguishable by having a top (upper side) and bottom (lower side), where more of its roots are produced. All tropical bamboos (e.g. the clump formers) have pachymorph rhizomes. Pachymorph rhizomes have short necks (e.g. *Bambusa, Dendrocalamus*, *Thyrsostachys*, *Gigantochloa*), but the species *Melocanna* sometimes has its elongated rhizomes (Fig. 2.5).

Leptomorph Rhizomes: Leptomorph rhizomes are cylindrical, slender, or sub-cylindrical in shape. They are usually narrower than the aerial culm. The internodes are long rather than broad, relatively uniform in length and symmetrical, hollow (rarely solid) with the narrow central lumen partitioned by a diaphragm, and nodes may or may not be raised or swollen (Rao et al. 1989; Zamora 1994). The node bears a single bud and a single row of roots, as typified by Phyllostachys. However, others buds are not available on their nodes, and roots may be sparse or absent.

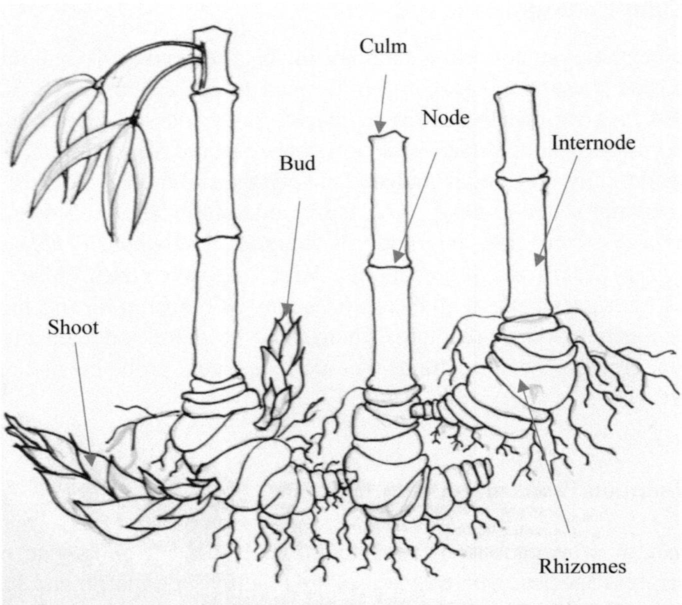

Fig. 2.5 Pachymorph (Clumpy) rhizome bamboo

These rhizomes are not distinguishable either by their bottom or top as most of the buds produce culms while some produce rhizomes. These types of rhizomes always have short necks. Some priority bamboos with leptomorph rhizomes are the non-clump forming species of Phyllostachys.

This method of propagation is applied whereby an offset or a rhizome of a culm from a clump is removed by digging around its rhizome and offsetting the rhizome from the main clumps. There are several ways to use leptomorph rhizomes with roots, and these include rhizomes with culm and roots and rhizome with culm-stock and roots. The rhizome with culm-stock and roots is similar to offset propagation, but there are a few differences (Fu and Banik 1996). This method of rhizome propagation involves separating culms from clumps and shortening them to the nodes above the culm, which is the most efficient method of reproducing young rhizomes into culms (Fig. 2.6).

Healthy rhizomes for propagation

- The rhizomes should be healthy and fresh in colour before being planted.
- They should not be more than two–three years old as those older than three years cannot produce high yield of buds, as well as shoots. When old rhizomes are used in the planting, the shoots cannot grow tall, leading to low-quality culms. The older the rhizomes, the less vigorous the production of shoots.

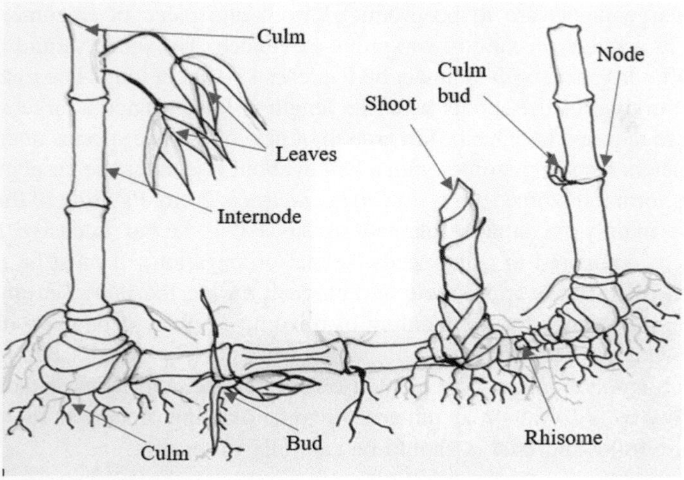

Fig. 2.6 Leptomorph (Running) rhizome bamboo

- When using rhizomes for this propagation, care must be taken with the rhizomes so that they are not damaged, as well as having enough roots around them.
- Rhizomes without culms are cut in a range of 50–60 cm long with about 10–15 nodes and with roots. In the process of transferring the rhizomes to distant places, care must take by wrapping them in moist peat moss or straw, after which they can then be covered with polyethene after the soil has been washed away.
- In this process of planting, some rhizomes need to be attached to the bearer's buds, while the rest are cut off.
- The young culms along with associated rhizomes and roots are used in this process. When the culm is young, its rhizome is maintained intact as compared to when it is large, since its top part can then be removed or the branches trimmed.
- This process is essentially for a rhizome with roots between 30 and 60 cm of the basal portion and connected to the culm only.

Post-Collecting Treatments of Rhizomes

The post-collecting treatment is when only rhizomes are used in the planting process. The rhizomes are collected and then laid horizontally to root for a period in a nursery. Another variation of this method is transplanting the rhizomes directly into the field or the farm after excavating them from a clump of bamboos. When laying them in the nursery bed to root out, the standard arrangement employs a ratio of 3:l soil with sand with the rhizomes planted 10–15-cm deep, the plants are separated by 25 cm (Banik (1997). Watering of the rhizomes is necessary, as well as a straw mulch to enhance growth. They should be kept under shade when young shoots emerge.

When large plants are to be produced from one piece of rhizome, they are separated in a range of one-to-two sprouts per piece. The piece should consist of three–four nodes each, with an intact bud at least 15–20-cm long. The rhizome may affect the survival of the shoots when its length is longer since a large amount of nutrients are required to enhance the growth of new buds. The species of culms with thick diameters require rhizomes with a length about five times the basal girth of the culm. The normal rhizome length for *Phyllostachys* is 2 cm. Planting of the bamboo rhizome is mainly for smaller plantations since it is labour intensive for larger farmlands as compared to using seeds (sexual propagation). It must be noted that planting of rhizomes is appropriate and efficient during the rainy season (Zamora 1994). The procedures for the planting of rhizome cuttings of bamboo directly on the farm is the same as that of planting stock from nursery raised seedlings, except that the cuttings are going to be planted directly in the field holes without putting them in Poly bags. To attain an efficient rhizome planting process in the plantation or farm, the following factors should be carefully observed:

- One must haul the offset cuttings to the planting site.
- One must loosen the soil in previously prepared planting holes.
- The cut rhizome is to be planted in a vertical position, with the lowest node of the culm offset above the ground.
- One must place the cutting at the centre of the planting hole and fill up the hole with soil, ensuring that the culms stand firmly in the holes.
- There should be regular watering of the plant, and mulching around the hole keeps the soil moist.

There is a limitation regarding its propagation as it is labour and time intensive removing the rhizomes. Hence the number of propagules to be extracted for larger farms or plantations will be limited. It is only appropriate for smaller farms.

2.2.4.4 Bamboo Propagation by Branch Cuttings

Planting of bamboo using branching patterns varies among several types of bamboos. The planting of the bud or culm node can potentially produce a branch, however, in many species, the original branch remains dominant and stout (Rao et al. 1989). In some others, buds are widely branched, while the original bamboo branch is uncertain from the other branches around. Each branch is a stem material. Primary stout branches have swollen bases, and this looks like the underground rhizome (usually referred to as a rhizomatous swelling).

Use of Branches instead of Culms

The process of planting bamboo through the use of branch cuttings is an effective method as the material can be handled very easily compared to bamboo culms. In this process, the thick-walled bamboo species with buds are ideal for planting (i.e. *Dendrocalamus* spp., *Bambusa* spp.). In this method, the likelihood of survival of the branch is only 50% assured, as in *D. hamiltonii* species. However, this method of planting bamboo or growing seedlings of bamboo is associated more with *D. asper* species. In most cases, it takes about 6–12 months nursing a branch plant to root and between 12–20 months for its rhizomes to develop the process of producing culms. Bamboo species that yield small branches at the top of the culm (e.g. *Thyrsostachys* oliveri and *T. siamensis)* cannot be adopted using this technique (Zamora 1994; Banik 1995).

Reduced Regeneration Time

It is necessary to reduce the regeneration time of branch cuttings in the field, since some branches may show natural aerial rooting with rhizome formation. The reduced regeneration is seen especially on injured culms and in congested clumps. When cuttings are taken from there, they are known as *pre-rooted and pre-rhizome branch cuttings*. Additionally, this method is can be used to stimulate growth of aerial roots and rhizomes near the base of branches. Regeneration time is significantly reduced when such materials are used.

2.2.4.5 Pre-rooting Propagation in Bamboos

Pre-rooting propagation of bamboo is a planting practice whereby the top part of the culm is chopped off or removed to allow for new culms to emerging from the clump. The latter, if done continuously for two years, is more efficient.

2.2.4.6 Pre-rooted, Rhizome, and Branched Propagation

In this process, the cuttings are prepared as follows:

- Roots should appear healthy.
- Branches should be selected from between 1 and 2 year-old culms.
- Branches should be cut from the culm using a saw. Using a cutlass or knives leads to the splitting of the culm.
- Branches that bear aerial roots, buds, and sometimes rhizomes, though not always, are selected.
- Well-drained and loamy soils kept under moist conditions are used to nurture the branches of bamboo.
- The branch cuttings are prepared by trimming leaves, small branches, and the branch tip with secateurs.
- The branch is trimmed between two–six nodes with fresh buds.
- The cuttings are planted into the nursery bed at two–three cm apart and the base at a depth of seven–ten cm. The sand formed around the base of the culm is then pressed firmly. Organic manure from animals and plants are used to cover up the exposed cut-off ends of the culm.

- The sprouting of the bamboo shoots occurs within a week or more. The roots of the new cutting will take between two–three weeks to start growing and as long as five–twelve weeks to fully develop.
- The culms start to develop immediately from 30 to the 60 days from the cuttings.

The cuttings should be taken from the bed before the new culms are produced and transplanted in Poly bags (15 × 23 cm) containing sandy 1oam and animal manure in a ratio of 3:l. The culm branches that are transplanted cuttings get hardened under shade for three–five days, but care is taken to make sure that the Polybags are well drained to prevent rotting of the culm cuttings.

2.2.4.7 Layering Bamboo

The layering of bamboo is a method of bringing a culm or branch in contact with soil so that propagation occurs. There are four methods of layering, namely: ground or simple layering; stump layering; air layering or marcotting; and seedling layering.

2.2.4.8 Ground Layering

The process of ground layering is whereby the whole bamboo culm or part of a branch of the culm is buried in a trench in the ground and enclosed with soil. In this case, young bamboo culms are used for layering. The top section of the culm is cut off to allow stimulation and growth of shoots. Branches and leaves on the upper part of the nodes of the culm, except for lower branches, must be kept at four–five nodes on each culm (Banik 2008). Bamboo species such as *Melocanna baccifera* with no main branches are trimmed into smaller branches. Species which can be used for this method include *Bambusa glaucescent, B. polymorpha, B. textiles, B. vulgaris, Guaduaangustfolia*, *Dendrocalamus giganteus*, and *D. long spathous*.

The main limitation associated with this approach is when the clumps are cultivated close to each other, making trench layers in between them difficult. This restriction of planting the culms too close to each other is so because the size and depth depend on the size of the culm being laid. The culm rooting medium used in ground layering involves the application of dry corn husks, dry leaves, dry coconut husks, and other dry weeds to cover the bamboo culm layering 5–9-cm deep in the ground. Regular watering is needed, but waterlogging must be avoided. The process of mulching around the culm will enhance the moisture of the soil.

2.2.4.9 Stump Layering

This process involves cutting the bamboo culm from the base of the node allowing for two–three basal nodes. Rooting materials such as dry corn husks, dry coconut husks, dry leaves, and other dry weeds are used to cover the stump on the ground. This approach is uncommon, and it was only reported to have been successful with the *Bambusa longispiculata* species. In this method, the stamps should only be covered by one–two cm of material. The application of 200 ppm IBA concentration is employed to treat the stump before covering it in the ground to produce shoots. However, sprouts usually emerge between 150 and 180 days after cutting the culms.

2.2.4.10 Seedling Layering

In this process of layering, the seedlings from the nursery bed must be 6–12 months old. The seedlings should be laid by placing them horizontally on the sand footing medium, and the culms are buried between 1.5 and 2.0-cm deep, but branches on the nodes are allowed to project above the ground (Banik 1995). Most seedlings at this age have three–four culms each with six–eight nodes. This technique is particularly effective for *Bambusa polymorpha* species, but shows low success with some other species, particularly B. *bambos,* B. *tulda,* and *Dendrocalamus strictus.* Usually, roots develop at the nodes in 60–90 days. The new shoots of plants are separated by cutting with secateurs and transplanting into Poly bags containing fertile soil.

2.2.5 Macro-Proliferation Methods

Macro-proliferation is the method of using cut bamboo rhizomes to produce multiple seedlings based on their ability to proliferate (Zamora 1994). The rhizomes to be nursed, when cut into pieces for planting or nursing, need to have roots and shoots, whereas single seedlings can produce multiples of shoots and seedlings ranging three–seven times depending on the species of bamboo. The method has been successful with seed producing bamboo species. Seeds are nursed in polybags as well as in nurseries immediately when they are collected from the field to produce the seedlings.

2.2.5.1 Raising Seedlings by Proliferation

Normally 15 × 23-cm polybags, containing soil, natural compost from decaying plants, or manure, as well as inorganic fertilisers, in the ratio of 3:1, can be applied.

The species that forms clumps produce about 48 culms over 150–270 days when they are ready for an increase.

2.2.5.2 Proliferating Seedlings

When the soil is washed from the root and rhizome system of the bamboo plant, then the old roots may be trimmed. Firstly, the rhizome is cut into pieces, each is replanted, and then hardened under shade for three–five days, while kept well-watered. After that, the transplanted pieces are brought to the nursery bed under the sun. The bamboo seedling can be multiplied in this technique during any month of the year, and its survival rate is 90–100%. As the young seedlings develop, they, in turn, can also be used as first seedlings from which new proliferating pieces can be produced once again (Zamora 1994; Rao et al. 1989).

This method is advantageous because, once seedlings are available, the process can be continued throughout the years of its survival. Proliferated seedlings are small in size, hence they yield a significant number of plants. The only disadvantage associated with this method is that, if the propagation continues for many years, then the multiplied plants might approach physiological maturity or flowering in later years. It is more advisable not to keep relying solely on this method of seed stocks, as there is a possible narrowing of the genetic base. Thus, different provenances or genotypes of the plant should be used, and, for each original seedling, the last multiplication should not exceed ten years (Fu and Banik 1996).

2.2.6 Post-care of Bamboo Cuttings

2.2.6.1 Poly Bags

The use of Polybags for potting of bamboo cuttings is, nowadays, considered standard practice. They are readily available and reasonably inexpensive for handling a large number of bamboo propagules. The polybags used for potting depending on the nature of the cutting and the species vary in size. Some common sizes and their uses are:

- Culm cutting: must have width 40 cm, height 50 cm, thickness 0.1 mm
- Branch cutting: must have width 15 cm, height 23 cm, thickness 0.1 mm
- Layered seedling: initially 10 × 15 cm, and 0.06 mm thick and finally 15 × 23 cm, and 0.06 mm thick

Bags usually used for potting are black. However, transparent Poly bags can also be utilised for the potting of culms. When bamboo culm cuttings are raised, it is not advisable to transplant them immediately in the same rainy season into the field. They can be prepared at least two-three months before the rainy season for well-developed rhizome systems to be produced. The cut culm seedlings in the Poly bags need to be in the nursery for about nine–ten months before they are planted and must be planted in a rainy season.

2.2.6.2 Selection of Land Site for Nurseries

In choosing a place for cultivation of bamboo and its product, one should take into consideration that its siting must be near where the seedlings are nursed in the seedbed for the seedlings to mature. The next sections treat further considerations.

2.2.6.3 Poly-bag Usage in Nursing Seedlings

Nursing of the seedlings of culm cutting in Poly bags should first be kept under shade for a number of days or weeks under partial sunlight in the 50–60% range. They can then be exposed gradually to full sunlight after a week. The polybag sizes employed in nursing seedlings of bamboo include large, medium, and small. The polybag usually are arranged in blocks of rows side by side with the bigger sizes in blocks four-bags wide; the medium and small sized ones are arranged in 10–12 and 15–17-bag wide blocks. However, a block with a width of 1.0–1.2 m is found to be ideal. Routine weeding and regular watering, as well as applying fertilisers, becomes difficult when blocks of seedlings in Poly bags are wider than the recommended 1.2 m dimension (Rao et al. 1989; Pattanaik et al. 2002). However, the block length can be adjusted to take into account the available space of land for the nursery. The adjustment of polybag arrangement can be 4–6-m blocks to minimise overcrowding and over shading. There should be 0.7–0.8-m wide footpaths of polybag arranged in blocks for accessibility by workers and for transport (Banik 1997; Midmore 2009).

2.2.6.4 Choice of Nursery Site and Nursery

The choice of nursery site for a bamboo seedling nursery is essential when propagation. As polybag are left for extended periods of time (i.e. six–eight months), cuttings produce strong rhizome systems and roots that can penetrate neighbouring polybags creating an intermingled rhizome and root mass of cuttings. The choice of site for a nursery is always a challenge when it comes to selection for propagation. It also creates a problem for the transportation of cuttings of culms from the nursery to the farm or site. Care must be taken during the separation process of the polybags not to disturb the roots, and the rhizomes in the Poly bags when shifted from one bed to another at three-month intervals and to minimise challenges. If any culm cutting starts producing flowers, it is to be marked. It must be noted that the flowering cuttings will die and therefore, should not be planted in the field (Rao et al. 1989).

2.2.6.5 Routine Activities Carried Out on Nursed Plants

The nursed bamboo plants require regular weeding, watering, and addition of soil to the bags, as and when needed, to help prevent washing of soil from the nursed plants as long as the cuttings are kept in the nursery beds. The few problems associated with bamboos such as pests, insects and animals attacks are treated with care in the nursery beds. Insects and pests which attack the nursed plants are sprayed with insecticides or pesticides,while the nursed areas are fenced to prevent wild and domestic animals from feeding on the plants' leaves and the shoots emerging from the ground (Banik 1994).

2.2.6.6 Field Performance of Various Planting Methods

The percentage of culm branch cuttings and macro-proliferation seedlings are superior to those of offsets and rhizomes. The size (height and diameter) of the fully-grown culms produced by offsets, rhizome, and culm cuttings are comparatively greater in the first year of plantation than those generated by branch cuttings. However, due to juvenility and the dynamic nature of the rhizome system of branch cuttings and proliferated seedlings, within three–four years, culms attain more or less similar sizes as those of offsets and rhizomes. Once the clump is well-developed (i.e. in five years), the seasons of planting, the originated branch cutting, or the offset becomes impossible to distinguish between (Banik 1995).

2.3 Pre-planting Operations

Several factors need to be considered before transporting bamboo seedlings propagated from seeds or vegetative stems or rhizome methods to the farm for transplanting include (Banik 1994, 1995; Fu and Banik 1996).

- Selection of land or planting site for the cultivation (type of soil, type of species to plant, climatic conditions).
- Preparation of land or planting site (clearing or weeding, burning of weeds, uprooting of stumps, levelling of land, etc.). The land site can be prepared using manual labour or mechanised methods.
- Propagation techniques or methods (seeds or vegetative propagation).

2.3.1 Selection of Land Site for the Cultivation of Bamboo

- Though bamboo species grow well in a wide range of soils under the right climatic conditions in an area having an altitude ranging from sea level to 1500 above sea level, it is important to critically look at the land or site selected for the cultivation. The land for a plantation must be fertile, because the quality of the soil enhances the growth speed of the bamboo. Hence, the site or land for the plantation is required to have fertile soil, which is moderately acidic, loamy soil, with a high-moisture content and preferably a pH of 5.5 to enhance the growing off well-developed quality culms (Banik 1994, 1995; Fu and Banik 1996).
- The loamy soil contains a good balance of clay, sand, and silt particles. Whenever the land is known to include soils that are too sandy or clayey, they can easily be modified using mulch and composts. Sandy soils drain too quickly and retain only a few nutrients, while clayey soils become waterlogged and trap nutrients in the ground, making them unavailable to the bamboo. Thus, soils can be improved by adding manure when sandy, thus providing drainage in the plantation for easy flow of water and also by adjusting the pH level.
- Tropical to moderately temperate climates are best suited for raising high-density plantations of bamboo seedlings.
- Selecting the type of species to plant at the land site is also necessary since some grow well in particular soils and geographical areas.

2.3.2 Preparation of Land or Planting Site

- After selecting the site or land to be used for the plantation of the bamboo, it is necessary to first weed or clear the land of all weeds and other plants on the site, either by manually or by mechanised means. If there are stumps on the land, they must be uprooted and removed as they may regenerate to compete with the growth of the bamboo seedlings (Banik 1994, 1995; Fu and Banik 1996; Pattanaik et al. 2002).
- The weeds and the other plants that have been cleared from the land are then burned on the ground once dry, especially in the dry season or allowed to decompose into the soil.
- The site is then levelled to prevent collection of rain or irrigation water on the plantation, particularly when stumps of trees and other plants have been removed. The land preparation can be done using manually or by mechanically.

2.4 Planting of Bamboo Seedlings

Planting of bamboo in the farm can be done using seedlings, culm cuttings, or offsets of rhizomes. When using seedlings nursed from seeds or culm cuttings that are grown in a nursery in the polybags, then it is necessary during planting to cut off the polybag to remove the mature seedlings before placing them into the dug pit. However, critical care needs to be taken in order not to disturb the rhizomes and the roots of the plant. After placing the plant in the pit, soil should be mound around the plants by pressing it with the feet.The surface of the pit after filling needs to be slightly sloped to one side to help maintain moisture around plants. In the case where a rhizome is being used as a seedling for a plantation, it must be noted that the technique is not different from the former. However, the depth of planting rhizome seedlings would be greater than that of the seeds and the culm cutting seedlings. Depending on the parent clump size, plant spacing of bamboo can range from 4×4 m to 10×10 m. Higher densities of spacing are appropriate for smaller sized bamboos, while lower densities require more space, which is suitable for larger bamboos (Zamora 1994; Banik 1994, 1995; Fu and Banik 1996; Pattanaik et al. 2002).

2.5 Post-planting Operations

Even though bamboo does not need much maintenance, to ensure its sustainability and healthy plantation, several post-planting activities must be carried out. Some of the post-planting processes require urgent attention during the early stages of the bamboo plantation while others are the subsequent steps. This maintenance process includes: replacement of dead seedlings; maintenance of culms and clumps; maintenance of soil; mounding; mulching; weeding; pest and insect control; pruning;, and so forth.

2.5.1 Replacement of Dead Plants

The newly transplanted offset cuttings and nursery seedlings will not all survive in their new environment. Hence, close monitoring is a critical requirement to ensure the survival of the plants. The dead seedlings (culm cuttings, rhizomes, and seedlings) that do not survive should immediately be replaced, and constant watering is essential for their survival.

2.5.2 Culm Maintenance

The maintenance culture of a plantation should be taken into serious consideration. Maintenance on the plantation is mostly done during the first two years after planting to protect the young plants from competing with other vegetation, pests, and rodents. After the first two years, maintenance activities are mainly confined to clump maintenance.

2.5.3 Maintenance of the Soil

The need to improve soil aeration around the plants to have healthy rhizome growth and shoot production is paramount, especially during the dry/summer season, and this is accomplished by keeping the soil moist. This practice also enhances the water retention capacity and the fertility of the soil. It is important to loosen the soil during the first year of a seedling plantation because this helps to correct the soil's air permeability, and temperature, and to decrease the number of weeds growing in the plantation that may compete with bamboo seedlings for nutrients and water, as well to improve the soil's chemical and physical profiles. Hence, this practice can be done about twice or thrice a year for each plant to enhance their growth. In the process of loosening the soil around the bamboo plant, care must be taken not to disturb or damage the rhizome system underneath (Fu and Banik 1996).

2.5.4 Weeding

The growth of bamboo plant is hampered by weeds which also compete for food and water with bamboo plants. These weeds compete with bamboo plants for nutrients, sunlight, and water. Constant weeding of the plantation is necessary to enhance the establishment of rhizomes and roots of the bamboo, as well as clumps. Weeds and stumps have rhizomes which sprout again if not removed (Fu and Banik 1996). It is critical to control and arrest the growth of weeds and other plants around each bamboo clump. When this is not done, then invariably it will result in poor root and stem development in the young bamboos plant. Weeds should be removed thoroughly, systematically, and regularly. It is critical to keep each bamboo plant clear of weeds and vegetation in a 0.6-m radius. Weeds should be removed when the soil is moist and also properly disposed of and burnt when dry. However, when the clumps of bamboos become well developed, they start to shed their leaves which prevent the growth of competing weeds and other plants within the plantation (Fu and Banik 1996).

2.5.5 *Pests, Insects, and Animal Protection*

The presence of parasites, insects, and grazing animals within the environment of the bamboo should be systematically controlled during the early stages of bamboo growth. In areas or communities where there is a likelihood of animals invading the plantation to graze on the plants, protection for the seedlings must be taken seriously to prevent the animals access to the farm and, by so doing, to maintain the quality growth of the seedlings to be produced. Fencing of the plantation is mostly associated with smaller plantations, while, for commercial farming, appropriate supervision is suggested to prevent grazing because fencing a large area is more expensive, considering the longer duration it might take for the bamboo seedlings to mature fully as culms. To monitor and control any defects or outbreak by insects and animals on the plantation, it is necessary to patrol the length and breadth of the farm to check each plant for damage, to assess the cause of damage, and then to find remedies to solve the problem. Bamboo is not easily attacked by pests and insects while on the plantation. However, those insects and pest that do attack it can easily be controlled by using appropriate pesticides and insecticides should there be an outbreak. Such untreated occurrences tend to hinder the growth and quality of the culms.

2.5.6 *Mulching*

This farming process has proved to be an effective means of improving the growth of bamboo plants in the plantation. Mulching is vital in plant cultivation, more so during the dry season, and bamboo needs a rainfall not less than 1000 mm. Growth is also enhanced by reducing evaporation of water from the soil when the plant is protected from direct sunlight and when the growth of weeds around the clumps of bamboo is checked. Branches and tips of harvested culms, sheaths, and leaves, which accumulate at the bottom of the bamboo clump and over time, serve as a manure to the plant. The process of uniformly spreading layers of leaves, branches from parent bamboo, and other organic material on the soil surface results in decomposition at the base of the culm that serves as mulching materials, as well as manure. When the natural decay of the mulch releases nutrients, it also does improves the soil moisture through organic carbon increment. This process is described as an efficient method to prevent rapid growth of weeds on the farm, as well as serve as a medium to check evaporation of excessive water from the soil under the plant. This process is a requirement that needs to be followed if high-quality bamboo seedlings are to be produced. Juvenile shoots require mulching because it provides protection from direct sunlight and ensures soil moisture, which ultimately results in the optimal growth of the bamboo without hardening and losing their good quality.

2.5.7 Soil Mounding

This method of maintenance on bamboo rhizomes and roots in clumps is usually done to prevent retardation in the growth of rhizomes when they are exposed in the soil to sunlight. Maintenance of clumps occurs when the clump is getting older, and the soil around the clump has been eroded, as the soil moisture tends to also be reduced, as well as the soil nutrients. Mostly these challenges occur when roots, emerging from the rhizomes and needing soil to grow sufficiently into culms, tend instead to be left to the mercy of the direct sunlight. The rhizomes and roots systems are critical parts of bamboo as they are the main networks that hold the tall and cumbersome culms in the ground. When the rhizomes are from sympodial bamboo, they grow horizontally from beneath the soil, while the new shoots grow upward from the soil. However, if these rhizomes and roots are not covered with soil, the culms may fall over to the ground if any strong wind or rainstorms should occur, since they lack enough soil to hold them. Soil mounding is usually done on older plantations of clump bamboos with the rhizomes being exposed from the soil. To prevent the tall and cumbersome culms from falling over at the occurrence of rain and storms, first old soil around the clump must be loosened, and then new soil should be mounded around the clumping bamboo to a height of about 20 to 60 cm, covering the base of the clumping bamboo.

2.5.8 Pruning Clump of Bamboo

This process involves removing large branches of culms and branches at the bottom of the clump to prevent congestion. If this is not done, it will affect the development of the new shoots. Pruning of bamboo branches can be done after the clump is three- years old and repeated every year. The thick and excess branches and culms of dried and dead branches, as well as other culms, are removed from the clump to allow for quality culm forming. Also, to permit access for new shoots to germinate, pruning and thinning facilitate growth.

2.5.9 Clump Management

The practice of proper management of bamboo clumps on a plantation or farm and attention to the sustenance of the plants foster high-quality yields of culms. Bamboo clump management provides not only high production yield of culms but also provides easy maintenance on the plantations of clumps on the farm. This system of management is necessary when culms grow close to each other in a clumpy area, making quality development of juvenal shoots difficult in some species. The clump management system is sometimes described as a partial a harvesting process, whereby unwanted culms areremoved from the class of clumps to prevent congestion in the plantation.

The maintenance of a bamboo clump is necessary for regeneration of new shoots of culms. Firstly, all unwanted branches need to to be thinned or pruned off from the interior sections of the clump, allowing it to form a "C shape" to facilitate and create an opening in the bamboo clumps (Fig. 2.7). Mostly, this can be accomplished only when a systematic management scheme is put in place to monitor the bamboo on the plantations. However, proper maintenance of clumps enables a 90% possibility of germination of new shoots, mostly at the outer boundaries of the clump, while the clump broadens in extent and the mature culms are dominant in the inner section. When cultivators of this resource plan for systematic management of clumps, sustainable development of bamboo and maximisation of productivity for a maximisation are fostered.

When no systems are created to manage the clumps on the plantations, the clumps tends to get overly congested, resulting in the deterioration of the bamboo quality, as well as quantity. It is critical to know that when clumps are congested, it is not easy to harvest culms because many young culms from the clump may end up being destroyed. When proper management of clumps is always practised in bamboo cultivation, harvesting of culms can be done without any difficulty, and both high quality and quantity of bamboo culms are produced. A management

Fig. 2.7 A class of clumpy bamboo to be managed through prunning

technique known as thinning, which is employed on a bamboo plantation, helps to provide space in a clump for the emergence of new buds. Practically, thinning involves sacrificing some culms in a clump to make possible better shoot production. It is also necessary to remove older culms, as well as dead and rotten culms, from the clump of bamboo to enable healthy growth of the younger shoots. Much attention needs to be placed on the already harvested culms that become rotten in the stubs and need to be removed from the clump by digging the rhizomes completely from the ground using axe or pickaxe. Also, it is important to make every attempt to extract any identified infected culm which is rotting from the group of clumps. Any further infectious disease should be reported to a pathologist who should suggest the required remedies and control measures to eliminate the disease.

2.5.10 Protection Bamboo from Fire

Bamboo seedlings, when planted on the farm, need to be protected from fires, which may occur from nearby bush or from within the farm. In the case where seedlings are well grown on the farm or the plantation with mature roots and rhizomes beneath the soil, it may not be too serious if a fire occurs. The mature culm may have its upper section destroyed without affecting the bottom of the plant that comprises the roots and rhizomes of the culms. However, young seedlings with immature roots and rhizomes may be highly affected by any outbreak of fire. The young bamboo plants' upper and lower sections, comprising the roots and the rhizomes, may all be destroyed during the fire. However, if the upper parts of the mature culms are destroyed, with the start of rains or with the provision of irrigation to the site to enhance the soil's moisture, the bamboo culms would easily sprout again. However, in the case where the roots and rhizomes systems are totally destroyed, there must be a replanting to replace the dead seedlings. It must be noted that the affected plant's growth and the quality of the regenerated bamboo will be affected.

2.6 Harvesting and Handling

Efficient management of bamboo clumps involves the systematic harvesting of mature culms from a group of clumps without affecting the environment of other culms, and this enhances continuous sustenance of production of culms. When this procedure is followed carefully on the plantations, it tends to increase the annual yield (Farrelly 1984). However, harvesting of bamboo culms on the plantation depends on the type of species that have been cultivated, since some species can be harvested after four years and others not until after six years. Harvesting from a clump of bamboo that has been scientifically managed through proper pruning, thinning, and fertilising does enhance annual harvesting of culms through proper management programs. A well-thinned clump from the early stages of planting gives a quality yield of culms, since spacing in a clump enhances efficiency in the harvesting of selected culms. The implementation of proper supervision of harvesting culms promotes regeneration of young shoots of culms based on which a constant supply of raw bamboo culms for industrial purposes is sustained.

The cutting off of the mature culms also helps in maintaining the vitality of the plant and also ensures germination and regeneration of new shoots. When harvesting mature culms, it is advisable to harvest culms which are older than three years, and only about 70–80% should be harvested, leaving about 20–30% of the mature culms to protect the young seedlings in the group of clumps. The reason for this practice is to ensure the young culms do not fall over as a result of high winds or storms, and this does help the growth. Culms that are in the age range of one–two years should be cared for during the harvesting of matured culms, along with a few of the three-year-old culms that are vigorous. This process would help the culms left in the clump to mature and also to regenerate young culms along with the already developed ones, after which, they may also be harvested. It must be noted that, as bamboo culms age, they deteriorate and eventually, die off and rot.

However, sustenance of bamboo clumps for regeneration mainly depends on healthy rhizomes systems from the class of clumps, which annually regenerates new shoots that develop into culms.

There are no records about the age of bamboos growing in natural forests, and due to this, it is impractical to use the difference during harvesting. However, bamboos that grow in their natural habitats regenerate and remain sustainable throughout their life cycle, whether harvested or not. Since harvesting of bamboo on a commercial farm or plantation depends on factors of species and age for its harvesting and handling, it is suggested that proper records need to be strictly maintained. It is recommended that a new plantation of bamboos should have its first harvesting six years after the time of cultivation. However, annual harvesting of culms can take effect after the first six years, or it can be determined within the ensuing years.

Sustenance of the culms in a clump is required for continuous harvesting, so therefore the implementation of an appropriate cutting of culms needs to be strategically implemented for quality bamboo plants. Bamboo sustainability integration is regarded as a mean of enhancing culm production and the life cycle of clumps, which largely depends on how efficiently the culms are selected during harvesting and how the extracted culm are processed into diverse products

Harvesting of bamboo culms from their natural habitats (forest) or a plantation is very laborious, and, because of this, it is suggested that efficient harvesting equipment need to be employed in this process to induce stress on those involved in the harvesting.

Suggested harvesting procedures to apply for bamboo culms include:

i. Bamboo culms selected in the harvesting process must be older than six years, if that is the first harvest since its cultivation, while for subsequent years it could be annually.

ii. Harvesting of selective culms from the clump require leaving an allowance on each culm ranging 0.15–1.3 m above the ground, or immediately after the first node above the from base. Cutting the culm to the given dimension and at the required point helps prevent stagnant water from accumulating in the internode as this could lead to insects breeding in them, as well as culms rotting.
iii. Harvesting of mature culms from a class of clumps is more appropriate done in the dry season of the year as compared to rainy season. It has been observed that starch content of culms reduces in the dry seasons, and, by this period, the culm is resistant to attacks by wood borers.
iv. Harvesting of culms from clumps needs to be carefully observed during the exercise in order not to cut the juvenal culms, as they might be fragile and the slightest touch of a sharp tool edge may easily destroy them.
v. Matured and healthy culms (three–five years) must be selected for harvesting.
vi. Culm harvesting should start from the central portion of the clumps since most of the matured culms are located in the inner sections.
vii. Harvesting of the culms must be performed with very sharp tools, and the harvesting tools should be disinfected using bleach as this can prevent any bacterial-risk infection to both the harvested and unharvested culms.
viii. Harvesting of quality and matured culms should not be above the juvenal culms in the group of clumps, as they need protection from the matured culms to shield them from a storm or the wind, so they do not collapse.
ix. Harvesting from a group of clumps requires leaving a sizeable number of culms in the groups of clumps for their sustenance. However, in an outbreak of disease in a clump, total cutting of culms is suggested to limit spreading to other clumps.
x. Harvesting of matured culms is recommended during the dry season of the year unless control of congestion from the clumps is necessary.
xi. Harvesting of culms from the class of clumps with the aim of preventing congestion and also maintaining high accessibility to culms in clumps should be executed with the technique of forming a C-shaped opening. This technique must make room at the periphery for the emergence of new shoots, resulting in the growth of multiples of culms.
xii. Harvested culms should be immersed in reservoirs filled with water or in rivers to aid leaching of starches and sugar from the culms. The leaching process provides resistance to insects and fungal attack on the culms, as these pests primarily feed on culm nutrients for survival. Storage of harvested culms in a river or tank can require take days or weeks for their preservation.
xiii. When there are no rivers or reservoirs of water at the site to treat them, they can be hauled to a flat-surface area where they can be air dried to aid in the reduction of biodegradation of the culms. The culms should be placed on the wall horizontally at an angle of 60% to allow for adequate ventilation to dry the culms.

xiv. The branches and leaves produced from the culms as waste material should be used as a mulching material on the remaining group of clumps, as it may serve as organic manure after its decay and hence enrich the soil.

2.7 Attacks on Harvested Bamboo

Immediately once bamboo culms are harvested, they are exposed to various forms of attack by pests, fungi, and insects since it is a weak material, and these pests, fungi, and insects can cause decay in a culm when it has not been treated on time (Mohanan 1997; Xu and Wang 2004). Some examples of fungi and insects include powder post and beetles, respectively. Bamboo culms lose quality as soon as fungi or insects attacks the culms, deteriorating its properties and applications value. There are different kinds of fungi and pests that may attack and destroy the entire culm if not properly treated. Also, when bamboo is harvested and the culms packed on each other without proper treatment, they become more vulnerable to these insects, fungi, and pests.

2.8 Insect Attacks

An insect is a small arthropod with three parts (head, thorax, and abdomen), having two pairs of wings and three pairs of legs. They are noticeably the most destructive pests that attack wood or bamboo culms (Wang et al. 1998). They usually survive well in the moist and warm tropics for reproduction or breeding. There have been several different groups of insects reported as attacking harvested woods, as well as bamboo culms and their products. Two of the principle types are wood borers and termites. The wood borers can be classified as wood boring beetles, bostrychid, moths, cerambycids, horntail wasps, ambrosia beetles, and lyctids. They can be described as organic material recyclers, but they cause great damage to wood and its products because they chew the internal cells and leave the wood or the culm devalued in quality. The process starts as the larvae develop inside the wood and penetrate into the culm further, chewing the tissues inside the wood or the culm. The larvae, after some weeks or months, then change from pupa into borer beetles which continue feeding on the tissues until they reach the cortex, where they then exit through a hole created on the surface. Observing drops of pills from the hole tunnel on the surface of the culm is the only means by which one can detect that there is an ongoing attack on the bamboo culm. The main function of female adults beetles is to breed which does not itself destroy bamboo tissue (Liese and Kumar 2003).

2.8.1 Beetles

Powder-post beetles can destroy a stored culm of bamboo reducing the tissue to a flour-like powder, with the outer cell becoming thinner, over without being noticed. They easily enter the bamboo culm within a day or two of harvest and feed on the starch and the carbohydrates within the parenchyma cell of the culm. This insect does destroy the culm tissue, as they create tunnels through the tissues of the culm leaving only a thin hard cortex of surface, which is misleading when taking into consideration the extent of the damage caused. They usually exit the culm after shredding the internal starchy tissues of the culm through to the hard cortex via circular boreholes (Liese and Kumar 2003). The effects of beetle attacks on bamboo culm are influenced by seasonal factors because the larvae rely mostly on the existence of starch to survive, and this is affected by the season. Beetles easily attack young culms because of their high moisture content as compared to older culms with less moisture content, as well as starch (Liese and Kumar 2003). It has been reported that the intensity of beetle attacks on bamboo culms depends on the species, with flowering bamboo culms having low moisture content and starch being hardly attacked. This is because the culm utilises its starch for seed production and therefore has a high resistance to beetle attacks. Beetle hardly attack some flowering species such as Pleioblastus because of its low moisture and starch content. On the other hand, *Bambusa vulgaris* is easily attacked by these insects due to its high content of starch and moisture. Other known wood-boring beetles do demolish the culms of bamboo by creating bigger holes inside the culm cells; these beetles includes Lyctus spp., Dinoderus spp., Wasps, and Carpenter bees which can seriously destroy bamboo and its products.

2.8.2 Termites

This type of insect can be characterized as silent destructive ants that cause chaos to wood and fabrics made from wood, as well as other products made from plants, not excepting bamboo. They are said to have wings which they shed upon discovering a habitat and also have the trait of eating, throughout their lifespan, for 24 h a day. They are made up of several different species, with the most commonly known species relevant to this discussion including subterranean termites, damp-wood termites, dry-wood termites, and Formosan termites. The subterranean and Formosan species of termites are soil-dwelling species while dry-wood termites can live in wood or bamboo culms without contact to the ground and feed on the cellulose. This they can doas they do not require moisture for their survival, unlike the damp wood termites which require much moisture.

Subterranean termites are described as the most destructive species of termites which live underground and roam above ground in search of food; they also need high humidity and water to build their nest's as large mounds (Liese and Kumar 2003).

They can form galleries very fast, and, through these, they search for food, reproducing continuously so that some group of offspring flys off to start new colonies elsewhere. They consist of workers, soldier, and a reproductive group which feeds on all products made from plants, including bamboo part. Because their destruction can cause very costly damage regarding to bamboo culms and products, it is important not to allow the culms to have direct contact with the ground. Also one must make sure the culms are not stacked close to water sources, and, when any discovery of galleries leading to their home is made, it must be destroyed.

Dry-wood termites, on the other hand, create their home mostly inside the wood of plants including bamboo culms and other wood products which they feed on. This type of termite does not require moisture or ground for their survival. Their invasion of the wood or bamboo culms and its products is typically initiated by the flying adults, who penetrate through openings and then create nests by dig tunnels in the culm. They can easily destroy culms, even those used in constructions above ground level, and can only be detected at an advanced stage of its destruction. Dry-wood termites recycle their body moisture while feeding on the culm and can survive on the nominal quantity of humidity found in the culms of the bamboo (Liese and Kumar 2003). It is suggested to treat bamboo culms with chemical before their application, which may prevent this insect from getting near where they are stored or used for products, as well as to seal all openings or cracks on the culms and in the perimeters of storage rooms.

Damp-wood termites are similar in form to dry-wood termites but a bit larger as compared to others. Most damp-wood insects require high moisture content for their survival, and, due to this reason, they feed on the moist woods as well as group in dying woods or finds their way into bamboo culms that are wet after harvesting. Given this, it is advisable always to keep environments in which bamboo culms are stored dry to prevent damp-wood termites from attacking them.

Damage caused by termites on wood and wood products can be very expensive. Hence it needs to be prevented on bamboo culms and its products too. Attacks by termites on wood and its products are caused by the working groups, which create tunnels in the wood to look for food to feed the queen. They usually leave thinly walled layers after attacks on the culms of bamboo which leads to fast disintegration.

2.9 Fungal Attacks

Fungi can be described as a classified group of living organisms. They are not animals, plants, or bacteria. Fungal attackes are characterised by staining and decay of bamboo culms (Zhou et al. 2010). There are different types of fungai that operate under different environmental conditions.. Bamboo culms are easily infested by fungi when not treated well to provide resistance to attacks from fungi (Xu et al. 2006, 2007). They use the cell walls of the culm as their source of energy for their survival by feeding on the organic components in their contents (moulds and blue stain fungi). Fungi are microorganisms that are sexually propagated from bodies of fruits or through asexual reproduction. They are universal and develop under wet environments consisting of a vast number of hyphae. Fungi have a branching filamentous structure possessing a primary mode of vegetative growth in which a network ifs formed that together is called Mycelium, a result of enzymatic decomposition. The consequence of decomposition on the culm wall is mostly realised, at a late stage, as a discolouration when it already has been destroyed, and physical damage has occurred to the culm.

Mostly, mould fungi breed on the surfaces of the harvested bamboo. Decay fungi, which include brown-rot and white-rot, do live inside the culm of the bamboo breeding within the substrate, while others also live on the surface of the culm where the environment is moist. These bacteria can live anywhere globally under those conditions that promote their development, as well as enzymatic degradation.

Though fungal organisms are present throughout the globe, they can only grow under certain moist conditions. Fungal attack on bamboo culm needs to be prevented, and it is important to dry the culm below the fibre saturation, as well as 20% below the moisture content, to enhance resistance to any fungi attack. However, the material moisture content must range from oven-dry mass in the range of 40–80%. These fungi can mostly be found in bamboo growing areas where the temperature range is suitable for fungal operations. Furthermore, when mycelium is exposed directly to the sun at temperatures above 55 °C, it results in the ruin of the enzyme system by protein denaturation, while other decay fungi have mycelia that can survive for hours in the temperature range of 95 °C (Schmidt 2006; Wei 2014).

2.9.1 Blue-Stain Fungi

These type of fungi lives and grow well in a humid environment and belong to the group of a fungi class known as the lower fungi, Deuteromycetes. They usually enter the bamboo culm at either end of the cut sections, while the culm is fresh and moist, as they pierce the parenchyma cells. These fungi normally are found living inside the freshly harvested culms because of their high relative humidity, moisture which is above 70%. Blue-stain fungi grow by feeding on nutrients from the culms, and this includes carbohydrates and sugar, as well as starch stored in the parenchyma cells (Liese and Kumar 2003). However, the internal tissue of the culm becomes blue-greyish black in colour as a result of the pigmented hyphae produced by the fungi. The effect of the discolouration of the internal tissues of the culm also spreads to the surface of the culms under shade, leaving spots or streaks in a uniformly distributed form. The effect on the culm of the infection of this fungi is that the aesthetic property diminishes, mostly resulting in culms with split and flakes.

2.9.2 *Mould Fungi*

These type of fungi lives and grow well in a humid environment. They usually live on the surface of bamboo culms mostly found on both cross edges of the green harvested culms that possess a high relative humidity above 70%. The high amount of sugar nutrient in the culm, as well as impurities on the surface of bamboo culm, help the hyphae to feed on them without penetrating into the culm. They also feed on the nutrients from the culms, while they develop in the inner section which is exposed at the cut ends, as well as dwelling in some finished bamboo products which are transported from one point to the other in containers. Moulds do not influence the physical properties of bamboo culms; however, their aesthetic appearance gets reduced by these fungi (Tang 2013). There are great varieties of mould fungi which produce huge numbers of microorganisms of many different colours, including black and blue. These organisms, when exposed in an indoor environment, can cause skin irritations as well as respiration difficulties to humans. Mould fungi produce some mycotoxins substances which are poisonous to people's health.

2.9.3 *Decay Fungi*

These type of fungi can penetrate deeply into the culm tissue using their hyphae, and they belong to the group of a fungi class known as the higher fungi, the Ascomycetes and Basidiomycetes which are actually bamboo-destroying fungi. The Basidiomycetes lives in the lumen of the bamboo cells as they feed on it, grow, and produce various enzymes depending on the type of fungi that disperse into the cell wall. The enzymes may then decay elements such as cellulose, hemicellulose, and the lignin as the culm turns into white-rot type or may only deteriorate the cellulose and the hemicellulose without affecting the lignin. But the culm colour finally turns brown as it decays. The white-rot fungi are the major destroyers of bamboo culms. (Ashaari and Mamat 2000). The decrease in culms' strength is the result of enzymatic degradation due to the damage of the cell-wall elements. Deterioration in the colour of culms or weight changes are evident, and the strength properties of the culm might have decreased when subjected to bending test. The effects of weight reduction and a decrease in mechanical strength in culms lead to limitations in its application more than in the stems of timber, because the weight loss in bamboo directly lessens the strength of the fibres. Mostly, the initial decay of culms is ignored but later leads to high costs linked to the diminished safety and the maintenance of affected elements of the construction (Ashaari and Mamat 2000).

The other form of fungi is in the group of Ascomycetes, which is the soft-rot fungi. These fungi hyphae live and grow inside the cell wall of the bamboo and require a smaller amount of oxygen to survive by enduring higher moisture content up to 80% where other fungi cannot survive. They normally produce some form of change of the lignin molecules, consuming the cellulose or the hemicellulose walls elements, and have a high resistance to various forms of toxic concentrated chemicals used as preservatives. The infested culm gives no sign of surface mycelium as the colour of the culm turns from it natural cream yellow to a dark brown-black. (Liese and Kumar 2003). When culms are left on the floor, they are easily destroyed by soft-rot fungal, as well as termites.

2.10 Post-harvest Treatment and Preservation

Post-harvest treatment and preservation of bamboo culms are essential to its conservation and life expectancy. Because bamboo is prone to insects and fungal attacks, especially beetles and termites, as well as mould fungal infection, preservation treatment is required (Xu and Wang 2004; Mohanan 1997; Yu et al. 2011a, b). The attacks on the bamboo culms by these fungi and insects lead to the reduction of the durability of the culms and also their usefulness. It is for this reason that post-harvest treatments are required for bamboo to help mitigate the likelihood of danger of insects attack, thus extending its lifespan and increasing culm quality (Garcia 2005). Depending on the end use of a bamboo culm, a particular preservation technique is adopted for its processing before being applied for industrial utilisation.

The post-harvest treatment requires drying of the culms first to attain 12% moisture content, which is critical for any load bearing applications in manufacturing andarchitectural design work. The culms are also subjected to seasoning before the machining processes and finishing of the products to improve their durability and stability (Moran 2002; Liese and Kumar 2003). Two drying techniques used in the drying of bamboo to the required degree of moisture content are air-drying within a well-ventilated environment or in the sunlight and kiln drying.

2.10.1 Air-Drying of Bamboo

This method of culm drying is a traditional technique employed since ages into remove moisture content from the culm of bamboo by exposing it to the conditions of the environment. The air-drying process of bamboo culm is more economical as compared to the kiln-drying method. However, the challenge associated with air-drying of culms is that it takes longer than kiln drying, i.e., weeks through to months to reach the required moisture content. As a result of taking longer to become sufficiently dry, there is a likelihood of the culms being attacked and infected by fungi, especially the mould fungi.

The air-drying process only requires proper stacking for air circulation around the culms to dry them without the addition of any other heat apart from the ambient air. The two methods of air-drying bamboo culms are namely; horizontal and vertical stacking. In the case where culms are horizontally stacked to air-dry, it takes longer to dry, almost double the drying period than those which are vertically stacked. Bamboo drying by air depends solely on the weather to reach its required moisture content, and, since the weather cannot be regulated, the process of drying may take longer to achieve the humidity below 12%, as required for products processes (Montoya-Arango 2006).

Another approach used to enhance bamboo culm drying is by splitting bamboo culms into halves or strips to speed up drying operations (Fig. 2.8). The culm can also be air-dried under the shade with ambient air in a well-ventilated environment. Drying of large volumes of bamboo culms by air-drying requires tying of the culms

Fig. 2.8 Traditional air drying of bamboo strips in Ghana

into bundles and placing them alternately on each other to prevent bending. The drying of bamboo culms to lose their moisture content by air-drying and without the application of other heats requires several days and weeks to reach the moisture content below 12%. Another approach to air-drying of a green harvested culm bamboo is by placing them on the flat platform with heavy objects laid on them during the entire drying period to attained moisture content below 12%.

2.10.2 Kiln-Drying of Bamboo

This method of reducing moisture content in bamboo culms is an improved method over the air-drying process since it offers a shorter and more efficient means of drying culms to the required measurement. The process employs the use of a boiler or electrical power to inject heat into the chambers housing the packed culms. The culms are cut into the given dimension, either split or whole, before being arranged in the chamber for drying. The elemental conditions, including relative humidity, temperature and air circulation in the chamber, are kept constant throughout the drying process until the 12% measure of moisture content is obtained (Tang et al. 2013; Yosias 2002). It is a recommended technique for drying large numbers of culms where high demand for production and high-level bamboo quality are needed (Yosias 2002; Tang et al. 2013).

2.10.3 Drying Rate and Defects of Bamboo

The rate of drying in bamboo when compared to wood shows that bamboo takes a longer time to dry the same mass of culms. A bamboo culm is characterised by an anatomical shape consisting of a cylindrical culm with a hole inside, resulting in a prolongation of the process in culms (Kumar et al. 1994; Liese and Kumar 2003). Also, the drying rate of thin-walled culms is faster than thick-walled culms, because the wall thickness enhances and also controls the speed of drying. (Laxamana 1985; Tang et al. 2013). It was also observed that younger culms have a slower drying rate due to the high moisture content in them when compare with the mature culms' lower moisture content. Drying of culms occurs quite rapidly at the start and later slows down gradually as drying advances.

There are defects associated with the drying of bamboo. These defects could be encountered during or after drying. The most common defects include rupture of culm tissues, node cracks, cell collapse, surface cracks, splits in culm, and end cracks. The difficulty of not having a consistent moisture content within an individual culm can result in discolouration during kiln drying, which does affect the quality of the dried culm (Tang et al. 2013). The green bamboo culms tend to be vulnerable to collapsing, due to tension in the vessels during the process of drying. Bamboo cells containing liquid collapse, leading to significant shrinkage in the cell walls (Tang et al. 2013; Liese and Kumar 2003). When bamboo culms are dried under low temperature conditions, the cell walls do not collapse but instead uniformly shrink in the radial and tangential directions. Different bamboo species require different drying temperatures, as well as varying drying rates since their drying characteristics are dependent on the wall thickness (Liese and Kumar 2003).

2.11 Preservation of Bamboo Culm

From various studies, it has been realised that culms of bamboos as natural materials are susceptible to natural and physical deterioration when harvested, especially at the young stages. The weakening of the bamboo culm is mainly a result of attacks by beetles, decay caused by staining fungi, and powder-post beetles, as well as termites. In an attempt to increase the bamboo-culm durability and serviceability, as well as its lifespan, there have been numerous techniques of conservation methods that have been introduced in the areas where they are cultivated. These methods used to extend the lifespan of bamboo culms are comprehensively classified into chemical and non-chemical methods.

2.11.1 Non-chemical Techniques

There are diverse bamboo treatments based on this technique that are less expensive but do not cause the reduction of carbohydrates, starch, and sugar content in bamboo culms. This treatment process has the advantages for the bamboo culm because most of the nutrients get removed during the treatment process and enhance its resistance to attacks by decay and mould fungi, as well as insects including beetles and many others. Below are some non-chemical treatment processes briefly discussed.

Curing This method involves the harvesting of the culms by cutting them from the base. After harvesting, the culms are allowed to remain together with the branches and leaves at the clump, as the respiration process continues to function and the starch and the sugar content in the culm reduces. This technique, when appropriately applied, will help prevent attacks by insects, but the possibility of fungi and termites' attacks cannot be eliminated entirely. Curing also helps improve the life span and quality of the bamboo culms.

Waterlogging More simple techniques used to treat green bamboos harvested from a plantation exist. The green harvested culms are soaked in water (running or standing water) for weeks or months to leach out the starch and carbohydrate content from the culm. The treatment of the culm by this process gives a high resistance to the bamboo culm against attacks by insects such as borers, but mostly there is a limitation to its resistance when it comes to dry-wood termites and mould or decay fungi (Ashaari and Mamat 2000; Nguyen 2002). Globally, these methods of preservation of bamboo culms are still being employed where the plants are distributed, more especially in Asia and some African countries. The method is commonly used in rural communities in Africa and Asia where bamboo resources are common. Hence, this is still a common practice used for treating bamboo culms for use in the construction of houses, handicraft, architecture works, and furniture. The waterlogging process is a comparatively cheap method of preserving the culm, and traditional artisans mostly employ this technique in the rural areas for their handicraft works. A study on the water treatment of bamboo species such as *Bambua vulgaris and Dendrocalamus strictus* was recently conducted in a village in Ghana. During the treatment process, the logged bamboos were left in a stream of water for eight weeks (Fig. 2.9). From the studies, it was observed that, after comparing it with the untreated culms which were also left for the same period of months in the seasoning process, the waterlogged culms had a higher resistance to insect and decay than the untreated. The result of the waterlogging treatment process can partially be compared to chemical treatment, though the life expectancy cannot provide for long-term preservation. This technique of preservation is only adequate for short-term products and therefore should be integrated into the treatment of long-term preservation technologies in order to reduce cost, as well as provide a high-quality resistance to insect attacks (Kaur et al. 2013). However, it was observed that the submerged culms in the running water after some time produce bad odours, as a result of bacteria activity on the culms, so this is one of the main defects associated with this the waterlogging process.

Fig. 2.9 Waterlogging treatment of *Bambusa vulgaris* species

Construction methods This method of using bamboo for construction purposes has been in existence for centuries, consistent with an appropriate selection and use of an effective method for construction to prevent damage over time. If the moisture content of a bamboo culm is less than that which makes it prone to insect and fungal attacks, and is kept below the fibre saturation point, then it can be used for construction. There exists a long-standing technique for treating these culms, without any application of chemicals as well as an efficient technique for construction suitable to be employed by many designers and architect (Janssen 2000; Amada and Untao 2001; Lobovikov et al. 2007; Heinsdorff 2010; Sharma and Gatóo 2014). The common practices include the placement of culms of bamboos on the foundations of walls or stones, which have been laid as preformed concrete footings for structures, rather than allowing them to have direct contact with the ground (Sen et al. 2011). It is advisable to harvest bamboo culms below a node base for firmness. It is also important to have continuous ventilation throughout the structure (Janssen 2000; Heinsdorff 2010).

Roofing overhanging will shield bamboo mats from the rainwater, and, usually, bamboo culms used for roofing are halved to foster run-off during of rain. They have a water-repellent cortex that improves their lifespan for several years until fruit bodies initiate internal degradation (Janssen 2000). Bamboo culms are used traditionally by experienced artisans for stable structures with appropriate roofing techniques. This method of preservation is very efficient, with high resistance to fungal and partial resistance against insects including beetles, but dry-wood termites can penetrate at certain stages. However, anytime subterranean termites attack a structure, it is recommended to remove their earthen tunnels and then use soil poisoning barriers to destroy their foundation (Nguyen 2002). With this method of treatment, bamboo culms sometimes are treated with boron against beetles, over four-to-five days, by piercing the internodal walls for complete diffusion (Heinsdorff 2010; Gutiérrez 2000).

Plastering is one of the oldest methods commonly practised in rural villages to protect the bamboo culms from being attack by fungi and insects. The process simply involves using whole, split, or mat bamboo for structures and then covering it with mud mixed with lime or any other organic materials for solidity. This method has often been employed globally bamboo resources are widely distributed. After making the framework of the structure, mud or the clay is then used to cover both sides of the structure's walls (Gutiérrez 2000). The purpose of the mud or cement technique being used, on the whole, split, or mat, bamboo panels, is to completely seal it. The bonding of the culms by the cement or the mud prevents raindrops from making direct contact with the bamboo material, preserving it from rotting, as well as from attacks by insects and fungi. Houses built with bamboo that has been sealed in cement or mud and have their surfaces finished with a lime wash have survived for decades (Jayanetti 2005; Liese and Kumar 2003).

Smoking This method of treating bamboo has been in existence for centuries. The technique is employed by stacking freshly harvested bamboo culm at the top of a building or kitchen above a fire source toe enable the smoke to penetrate into the culms tissues. This method is mostly practised in rural areas on freshly harvested culms to preserve them from attacks by insects. The treatment of the culm with smoke (carbon) forms a layer of protection on the culm and helps to reduce moisture content, thus providing high resistance to attacks by insects and fungi. The continuous interaction between the smoke from the fire and the heat leads to darkening of the colour of the culms. The technique has been adopted in the conservation of green bamboo culms as they are stacked in a furnace with temperatures ranging 120–150 °C. The process of producing soot, as well as pyrolytic chemicals on the surface of the culms, basically protect the culms against beetles (Leithoff and Peek 2001; Liese and Kumar 2003). During this treatment process, a strong chemical, which is acidic in nature, is infused into the culms sitting in the furnace, and this has a positive effect on their usefulness. These treated culms are basically for external uses, and, due to the chemical changes in the carbohydrates, they have high resistance to insects. Culms that are processed by smoking, when exposed to direct contact with the ground for a long period, will be vulnerable to attack by fungi and termites (Leithoff and Peek 2001). Smoke drying also reduces splitting of the bamboo culms (Liese and Kumar 2003).

2.11.2 Chemical Methods

This process involves the use of various types of chemicals that are efficient in expanding the durability of treating the culms , as compared to non-chemical techniques. This method of treatment is more expensive than non-chemical treatments. Hence, its application aligns with more vigorous treatments where a higher valued product is required. The following are some of the chemical methods which are employed in the chemical preservation of bamboo.

Butt Treatment This method of preservation is employed on both dried and freshly harvested bamboo culms immersed in a tank containing a preservative for their treatment. The bamboo culms are put into the tank or container for about one week. A Ghanaian company which manufactures furniture and other bamboo products employs this method of treatment of material for their products. They use a mixture of a chemical called Dursban and engine oil to treat the culms and so preserve them from insect attacks (Fig. 2.10).

Old Engine Oil The process of preservation involves the application of used engine oils to the green or dried bamboo culms, particularly those produced by rural farmers. However, this method of treatment has not scientifically been proven to be effective.

Steeping or Sap Displacement In this method, the chemical is poured into a basin or tank after which the culms are then positioned vertically in the reservoir

Fig. 2.10 Traditional Ghanaian method of preserving bamboo culms

and left for some days to the allow penetration of the chemical into the culms. This process is applicable to both freshly harvested bamboos and dried culms.

The Open-Tank Treatment This process of preservation involves the use of a water-soluble solution whereby bamboo culms cut into given dimensions, and then immersed in the chemical solution for days or weeks, experience adequate penetration of the chemical. The diffusion process takes place in all directions within the culms, with the highest rate being at the top and bottom end of the culms.

2.12 Treatability

The anatomy and moisture content of bamboo makes its treatment process different from normal wood and critical for selection of suitable preservation techniques. The culm outer layer is covered with cuticula which prevents solution penetration. This makes the use of simple treatment processes such as only soaking inadequate. The sclerotic parenchyma cells on the inner layer of the culm also hinder to some extent, penetration (Liese and Schmitt 2006). Unlike wood which has ray cells aiding in the uptake of preservatives, bamboo lacks these cells and relies on only the metaxylem vessels for preservative uptake. These vessels, however, constitute only 6–8% of the total volume of fibres. The remaining tissues, namely the fibre bundle and parenchyma must be treated by diffusion.

2.13 Preservatives

Bamboo culm needs a higher degree of antibacterial preparation as compared to wood. There are various ways by which bamboo culms can be preserved, and this includes the water-solution process, as well as chemical preparations needed for particular applications. The choice of preservatives is determined by the type of material being treated, as well as the life expectancy for particular products. These chemicals are either in the form of pastes or liquids, as well as tablets. The use of salt as one of the preservatives currently minimises the risk of hazardous chemicals being released into the environment by other preservatives, as well as consequent health problems. The application of water-solution preservatives has more advantages over other chemicals because they appear to be clean and odourless and more efficiently applied, as well as being less expensive (Liese and Kumar 2003; Liese and Schmitt 2006).

2.13.1 Waterborne Types

In this method of preservation, a salt solution is used to treat the bamboo culms by immersing them in the given solution with the aid of a weighting element. The water is allowed to evaporate, and the salt then remains inside the culms. This treatment process is categorised into two main types, i.e. the fixing and non-fixing method of bamboo tissues.

Non-fixing Types In this method, the application of preservative chemicals is mainly applied on the bamboo during the drying season for effectiveness, because, during the wet season, its efficiency might be reduced due to rains washing the chemical off. (Liese and Kumar 2003; Liese and Schmitt 2006).

Fixing-type In this method of preservation, chemicals bound in woody tissues both inside and outside are applied. This method is more useful for high and efficient preservation of bamboo culms.

2.13.2 Organic Acids

This form of preservation of bamboo culms is used to prevent attacks by moulding fungi on the freshly harvested bamboo. This preservation is only applicable for the prevention of mould forming on a freshly harvested bamboo culms, but only for short-term protection of the bamboo due to their high moisture content, during transportation from one point to another. The technique requires immersing the culms in a 10% acetic acid solution which especially fights against the growth of mould during the raining season (Tang et al. 2012).

2.14 Preservatives and Treatment

Bamboo treatment is very essential for the enhancement of its properties, so as to withstand all external attacks and also to improve its resistance to fungi and insect attack. The treatment processes with preservatives are categorised into two major categories, namely, pressure and non-pressure processes.

2.14.1 Pressure Treatments

This method of bamboo treatment is described as an efficient technique used to treat bamboo culms against conditions that are hostile to their quality. Pressure treatment accounts for deeper and even distribution of the chemicals through the entire culm, as well as retention of preservatives in the matter of the culm (Liese and Kumar 2003). This method is performed by forcing the preservatives into the bamboo tissue, by increasing the vacuum and the pressure upon the preservative in the treatment cylinder (Tang 2009).

The skin of the bamboo culm makes it difficult to accomplish this process, especially in the radial direction. However, to enable smoother penetration within the culms, holes on opposite sides of the lower and upper part of each internode must be made by drilling throughout the entire culm.

2.14.2 Non-pressure Treatment

There are various processes of non-pressure treatments, which include hot and cold treatment, soaking or diffusion, steeping or butt-end treatment, and sap-replacement processes. Two of the process have already been discussed briefly in this book.

Soaking/diffusion This treatment method of bamboo culms involves splitting green harvested culms into the required size and then after they soak in a water solution, chemicals enable diffusion to take place within the tissues of the culm (Liese and Kumar 2003). However, because specific gravity is less than 1.0 g/cm^3 in bamboo, it is advisable to bind the strips with a weight to help immerse them in the basin or reservoir before adding the chemical solution (Tang and Liese 2011). The diffusion process takes place in the axial direction, less in the transverse and slightly better radially than tangentially.

Steeping or Butt-end Treatment This treatment method of bamboo requires that green harvested culms be split into the required size and then, after soaking in a water solution of chemicals, for diffusion to take place within the tissues of the culm (Liese and Kumar 2003). However, because specific gravity is less than 1.0 g/cm^3 in bamboo, it is advisable to bind the strips with a weight to immersed them the basin or reservoir before adding the chemical solution (Tang and Liese 2011). The diffusion process takes place in the axial direction, less in the transverse and slightly better radially than tangentially.

2.15 Sustainability of Bamboo and Its Products

The term sustainability of bamboo and its products is very important as there is a rising awareness of sustainable development on a global scale. The global and industrial concern in recent years is about attaining the ultimate goal of sustainability (Finkbeiner et al. 2010; Klöpffer and Renner 2009; Klöpffer 2003). The need for sustainable development of bamboo is urgent because consumption patterns of bamboo and its products have continuously increased in recent years. The multipurpose functions of bamboo have led to rising pressure from many industries for its application, especially in construction, materials, and manufacturing sector as well as other new areas of use (Finkbeiner et al. 2008; Schau et al. 2012).

Bamboo is a fast-growing plant; however, its sustainability is at high risk and hence the need for its renewability to meet current and future demands for the engineering industry and other uses (Schau et al. 2012). Bamboo sustainability is imperative. A look at the stages of its life cycle reveals a versatile plant with a short growth cycle that can be harvested in three to five years, as opposed to the ten to 50 years required for most soft and hardwoods. It grows in almost any climate, and it can replace itself very quickly (Gratani et al. 2008; Van der Lugt et al. 2009). The bamboo plant has been used for thousands of years and has not been depleted from the forests. However, the introduction of modern and advanced methods that affects its life cycle need to be critically examined in order not to ignore any negative consequences to meeting challenges to its sustainability. Bamboo cultivation and production is a far more environmentally safe and sustainable option than other modern industrial materials like corrugated metal, artificial plaster, and chemically treated wood (Van der Lugt et al. 2009).

References

Amanda S, Untao S (2001) Fracture properties of bamboo. Compos B 32:451–459

Ashaari Z, Mamat N (2000) Traditional treatment of Malaysian bamboos: resistance towards white rot fungus and durability in service. Pak J Biol Sci 3:1453–1458

Banik RL (1991) Biology and propagation of bamboos of Bangladesh. A Ph.D. thesis, University of Dhaka, pp 1–321

Banik RL (1994) Studies on seed germination, seedling growth and nursery management of Melocanna baccifera (Roxb.) Kurz. In: Proceedings of the 4th international bamboo workshop on bamboo in Asia and the Pacific, 27–30 Nov 1991. FORSPA Publication No 6 IDRC FAOUNDP, Chiangmai, Thailand, pp 113–119

Banik RL (1995) A manual of vegetative propagation of bamboos. INBAR technical report no 6, New Delhi, India, pp 1–66
Banik RL (1997) Growth response of bamboo seedlings under different light conditions at nursery stage. Bangladesh J For Sci 26(2):13–18
Banik RL (2008) Issues on production of bamboo planting materials—Lessons and Strategies. Indian For (Bamb Iss) 134(3):291–304
Farrelly D (1984) The Benefits of Bamboo. The Sciences 24(6):11-12
Finkbeiner M, Reimann K, Ackermann R (2008) Life cycle sustainability assessment (LCSA) for products and processes. In: SETAC Europe 18th annual meeting, 25–29 May 2008, Warsaw, Poland
Finkbeiner M, Lehmann A, Schau EM, Traverso M (2010) Towards life cycle sustainability assessment. sustainability 2(10):3309–3322
Fu M, Banik RL (1996). Bamboo production system and their management. Bamboo production system and their management. In: Rao IVR, Widjaja E (eds) Bamboo, people and the environment. Propagation and management, vol I. Proceedings of the 5th international bamboo workshop and the 4th international bamboo congress, 19–22 June1995, Ubud Bali Indonesia. INBAR/EBF/IPGRI/IDRC, INBAR, New Delhi, pp 18–33
Garcia CM (2005) Management of powder-post beetles, *Dinoderus minutus* F. in freshly cut bamboo
Gratani L, Maria FC, Laura V, Guiseppe F, Digiulio E (2008) Growth pattern and photosynthesis activities of different bamboo species growing in a botanical garden of Rome. Flora 203:77–84
Gutiérrez JA (2000) Structural adequacy of traditional bamboo housing in Latin America. Technical report no. 19. INBAR, Beijing
Heinsdorff M (2010) The bamboo architecture design with nature. Hirmer, Munich
Janssen JJA (2000) Designing and Building with Bamboo. International Network for Bamboo and Rattan Technical Report No. 29. pp 207
Jayanetti L (2005) Seismic testing of a bamboo based building system. Presented at an INBAR international workshop in Beijing, China, 24–25 Nov 2005
Kaur P, Kardam V, Pant KK, Satya S, Naik SN (2013) Scientific investigation of traditional water leaching method for bamboo preservation. Bamboo science and culture. J Am Bamb Soc 23 (1):27–32
Klöpffer W (2003) Life-cycle based methods for sustainable product development. Int J Life Cycle Assess 8(3):157–159
Klöpffer W, Renner I (2009) Life-cycle based sustainability assessment of products, in environmental management accounting for cleaner production. In: Schaltegger S et al (eds) Springer Netherlands, Dordrecht, pp. 91–102
Kumar S, Shukla KS, Dev I, Dobriyal PB (1994) Bamboo preservation techniques: a review. Technical report no. 3. International Network for Bamboo and Rattan (INBAR), New Delhi, India; and Indian Council of Forestry Research and Education (ICFRE), Dehra Dun, India
Laxamana MG (1985) Drying of some Philippine commercial bamboos. J For Prod Res Dev Inst (Philippines) 14(1&2):8–19
Leithoff H, Peek RD (2001) Heat treatment of bamboo. International research group on wood preservation. IRG/WP/01-40216, Stockholm
Liese W, Kumar S (2003) Bamboo preservation compendium. International Network for Bamboo and Rattan (INBAR), Beijing
Liese W, Schmitt U (2006) Development and structure of the terminal layer in bamboo culms. Wood Sci Technol 40(1):4–15
Lobovikov M, Paudel S, Piazza M, Ren H, Wu J (2007) World Bamboo Resources: a thematic study prepared in the framework of the Global Forest Resources Assessment 2005. Food and Agriculture Organization (FAO) of the United Nations, Rome, Italy
Midmore DJ (2009) Overview of the ACIAR bamboo project outcomes. In: Midmore DJ (ed) Silvicultural management of bamboo in the Philippines and Australia for shoots and timber. In: Proceedings of a workshop Los Baños, Philippines, 22–23 Nov 2006. ACIAR Proceeding no. 129, pp 7–12

Mohanan C (1997) Diseases of bamboos in Asia. India Brill Academic
Montoya-Arango JA (2006) Trocknungsverfahren für die Bambusart Guadua Angustifolia under tropischen Bedingungen. Ph.D. Dissertation, University Hamburg
Moran JA (2002) Traditional preservation methods in Latin America. INBAR technical report no. 25, 70 pp
Nguyen TBN (2002) Investigation on preservation of bamboo used for construction. Ph.D. Dissertation, Forest Science Institute of Vietnam, Vietnam, pp 44–51
Pattanaik S, Pathak KC, Trivedi S (2002) Cohort mapping in bamboo resource management. In: Pattanaik S, Sing AN, Kundu M et al (eds) Proceedings of expert consultation on strategies for sustainable utilization of bamboo resources subsequent to gregarious flowering in the northeast, 24–25 Apr 2002. RFRI Jorhat, book no 12-2002. UNIDO, India, pp 11–17
Rao IVR, Yusoff AM, Rao AN (1989) Propagation of bamboo and rattan through tissue culture. IDRC Bamboo and Rattan Research Network, Singapore
Ramyarangsi S (1988) Techniques for seed storage of Thyrsostachys siamensis. In: Rao IVR
Schmidt O (2006) Wood and tree fungi. Biology, damage, protection, and use. Springer, Berlin
Schau EM, Chang YJ, Scheumann R, Finkbeiner M (2012) Manufactured products—how can their life cycle sustainability be measured? A case study of a bamboo bicycle. In: The 10th global conference on sustainable manufacturing, Istanbul, Turkey
Sen T, Jagannatha Reddy HN (2011) Application of sisal, bamboo, coir and jute natural composites in structural upgradation. Int J Innov Manage Technol
Sharma B, Gatóo A, Bock M, Mulligan H, Ramage M (2014) Engineered bamboo: state of the art. Proc Inst Civil Eng. doi:10.1680/coma.14.00020
Tang TKH, Liese W (2011) Pressure treatment of bamboo culms of three Vietnamese species by boron and CCB preservatives. J Bamboo Rattan 10(1&2):63–76
Tang TKH (2009) Bamboo preservation in Vietnam. International Research Group on Wood Protection. IRG/W/40457, pp 1–11
Tang TKH, Schmidt O, Liese W (2012) Protection of bamboo against mould using environmentfriendly chemicals. J Trop For Sci 24(2):285–290
Tang TKH (2013) Preservation and drying of commercial bamboo species of Vietnam. Ph.D. Dissertation, Hamburg University
Tang TKH, Welling J, LieseW (2013) Kiln drying for bamboo culm parts of the species Bambusa stenostachya, Dendrocalamus asper and Thyrsostachys siamensis. Journal of the Indian Academy of Wood Science 10(1):26-31
Van der Lugt P, Vogtländer J, Brezet H (2009) Bamboo, a sustainable solution for Western Europe Design Cases, LCAs and Land-use. INBAR technical report no. 30
Wang H, Varma RV, Tiansen X (1998) Insects of bamboos in Asia. An illustrated manual. INBAR, Beijing, technical report no. 13
Wei D (2014) Bamboo is inhabiting fungi and their damage to the substrate. Dissertation, Department of Biology, Hamburg University, Hamburg
Xu TS, Wang HJ (2004) Main pests of bamboo in China. Chinese Forest Publish House, Beijing, pp 45–46
Xu MQ, Dai YC, Fan SH, Jin LX, Lu Q, Tian GZ, Wang LF (2006) Records of bamboo diseases and the taxonomy of their pathogens in China(I). For Res 19(6):692–699
Xu MQ, Dai YC, Fan SH, Jin LX, Lu Q, Tian GZ, Wang LF (2007) Records of bamboo diseases and the taxonomy of their pathogens in China(II). For Res 21(1):45–52
Yosias G (2002) Preliminary study on the drying of bamboo (Bambusa blumeana) in a wood waste–fired kiln. Bamboo for sustainable development. In: Proceedings of the 5th international bamboo congress and the 6th international bamboo workshop, San Jose′, Costa Rica, 2–6 Nov 1998. INBAR 2002, New Delhi, India, pp 495–510
Yu DW, Tan HW, Ruan Y (2011a) A future bamboo-structure residential building prototype in China: life cycle assessment of energy use and carbon emission. Energy Build 43(10): 2638–2646

Yu HP, Shen K, Wang ZT, Mu LL, Li GQ (2011b) Population control of the yellow-spined bamboo locust, Ceracris kiangsu, using urine-borne chemical baits in bamboo forest. Entomol Exp Appl 138(1):71–76

Zamora AB (1994) Review on micropropagation research on bamboos. In: Constraints to production of bamboos and rattan. Report of a consultation held in Bangalore, 9–13 May 1994. INBAR technical report no. 5, New Delhi, pp 45–100

Zhou CL, Wu XQ, Ji J, Ye JR (2010) Research advances of bamboo diseases. China Sci Technol 24(5):8–13

Chapter 3
Properties of Bamboo

Abstract The properties of materials are the main elements that determine the appropriate applications of those materials. These properties also assist design engineers, architects, material scientists and other industrial users of materials to know where to apply them. A review of the physical properties, mechanical properties, and critical structural equations for analysis of bamboo structures are presented. Standardisation of testing procedures for bamboo is discussed and adapted for tensile and compressive testing. The tensile strength and modulus of elasticity of *Bambusa arundinacea* with nodes were found to be 114 and 126.2 MPa, respectively. The compressive strength of *Bambusa vulgaris* with and without nodes was found to be 113.54 and 98.80 MPa respectively. Bamboo treated with brine exhibited better tensile strength than samples treated with pure water or acid. Microstructure imaging is used to capture the behaviour of the vascular bundle when water and brine treatments have been performed.

3.1 Bamboo Growth Morphology

Bamboo is characteristically unique as it possesses tree- and grass-like features suitable for survival in many habitats and is often referred to as a natural composite. The physical features of bamboo are the branches, culm (hollow), flowers, leaves, rhizomes, sheath, and roots (Banik 2015; Resource Centre for Natural Fibre Crafts Project 2015). Morphologically, bamboos may be classified as clumpy (dense) or non-clumpy, with culms connected by joints called nodes. Bamboo growth is characterised by a primary shoot that grows in the longitudinal direction. Bamboos are generally classified as dwarf (short), tall, and climbing bamboos which thrive in marginally or fully deciduous forests (Banik 2015).

E.T. Akinlabi et al., *Bamboo*, DOI 10.1007/978-3-319-56808-9_3

3.2 Morphology of the Bamboo Plant

Bamboo plant morphology consists of branches, sheaths, culms, flowers, leaves, rhizomes, and roots which are briefly described in the following sections.

3.2.1 Culm

The culm is the most noticeable and visible part of the bamboo plant which develops from shoots that emerge from buds. It appears as an elongated cylindrical stem with intermittent joints known as nodes. Generally, culms vary in dimensions, colour and sometimes smell depending on the species. The culm has three main constituents, namely: stem, stem base, and stem petiole.

Variations in Culm Dimensions

There are major variations in the dimensions of culms when different species are compared. The culms may be erect, zigzag, bent, straggling, or clambering. The variations in diameter of bamboo with respect to species fall within the range of 1.0–20 mm. Banik (2000) reported that *Dendrocalamus strictus* had a maximum wall thickness of 3.45 cm, while *Schizostachyum dullooa* has the thinnest wall thickness of 1.2–6.5 mm. Banik (2015) observed that bamboo-culm diameter was a function of habitat conditions and age. As age increased, the diameter of newly generating culms also increased. The maximum culm wall thickness was located at the base and top regions, while the middle regions were reported to have the thinnest walls. The advantage of having thin walls is to aid in enhancing the elastic properties of the bamboo to withstand external loading from natural phenomenon such as storms.

Depending on the size of the culm, various degrees of the onset of taper are observed. For larger culms, an approximately cylindrical base (lower half) is reported, with the upper half showing significant taper. For smaller sized culms, a steady taper from the base along the entire length is observed (Banik 2015).

Colour Variations

The colours of culms for bamboo species span a variety of hues. While green is the most dominant colour, however, depending on the species, they may be yellow, brown, purple brown, striped, or black like *Phyllostachys nigra* (Fig. 3.1).

Culm Nodes and Internodes

The stem or culm of the bamboo plant may be conical or elliptical and comprises several nodes and internodes (Fig. 3.2). The internode is hollow with nodes at both ends (Schroder 2011). Two closely adjacent rings can be located above and beneath the node. The ring beneath the node is generally referred to as the sheath ring, resulting from scars of fallen sheath. The ring above the node is as a result of

Fig. 3.1 Variations in culm colours among species of bamboo plant

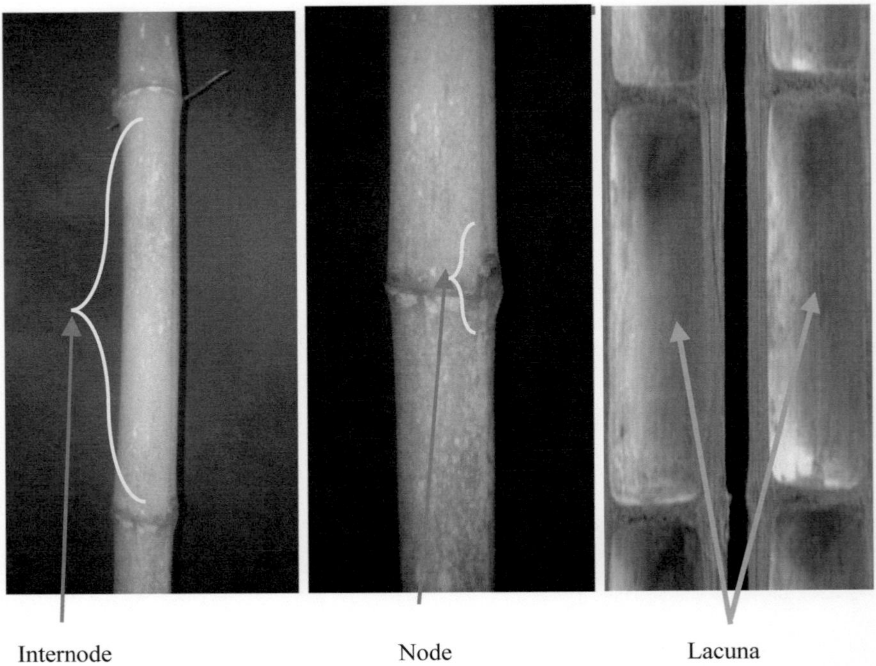

Fig. 3.2 Bamboo culm morphology

intermodal tissues ceasing to grow. The nodes contribute to ensuring that bamboo is relatively straight and enabling storage of water and nutrients (Liese and Ding 2003; Taylor et al. 2014). The internode length is a function of species and genetics, and therefore several variations are reported. Banik (2000) reports that the length of internodes and how they are distributed within the culm may be used as a factor in determining the habitat conditions of bamboos. Both culm sheaths and buds develop and emerge from the nodal region.

3.2.2 Sheath

The development of sheath is dependent on the sheath organ, which emerges from the node of the bamboo. Several types of sheaths develop in bamboos, namely: culm sheath, leaf sheath, branching sheath, and rhizome sheath. However, the most visible are the culm sheaths.

The culm sheath is basically a modified leaf comprising a blade, ligule, and auricles. The ligule is located within the inner part of the sheath along the portion where the blade connects. The auricle can be found on all sides of the blade base. There are huge variations in culm sheath colour, geometry, morphology, and

Fig. 3.3 Sheaths on culms of bamboo

location and placement in the nodal region (Banik 1997a, 2000, 2015). Chatterjee and Raizada (1963) reported the morphology is species specific for culm sheaths and is an important feature used for bamboo identification. Depending on the type of species, the sheaths may or may not fall off. Branch sheaths have layers that closely surround the branches emerging from the nodal region of the culm.

The size of sheaths gradually reduces from the base of the culm or branch to its apex (Fig. 3.3).

3.2.3 Branches

Branches develop from lateral sheath-covered buds, which emerge above the sheath scar from alternate sides of the culm. Branching can commence in the year when the culm emerges or the next year, termed syllepsis or prolepsis respectively. Syllepsis is observed in the *Bambusa genera*, while prolepsis is typical of *Melocanna baccifera*. Two major types of branching can occur, depending on the species namely: primary dominant-stout branching (Fig. 3.4) and numerous subequal branching. Stout branching is typical of the *B. genera*, which have pachmorph rhizomes, while numerous subequal branching is seen in *M. Baccifera* (Banik 2015).

Fig. 3.4 Branches of bamboo culms

3.2.4 Rhizomes

The rhizome of bamboo is an important subterranean organ or system with multiple branching that is responsible for the structure, stability, nutrient storage from photosynthesis, and territory colonisation (Fig. 3.5) (Long et al. 2003; Gross 2009; Banik 2015; Resource Centre for Natural Fibre Crafts Project 2015). Rhizomes vary in appearance (colour) and characteristics, depending on the bamboo species (Gross 2009). From research, two types of bamboo rhizomes are known, namely: Pachymorph (sympodial) and leptomorph (monopodial) (Table 3.1).

3.2.5 Roots

As with all plants, the bamboo root system aids in the transport and storage of nutrients, food, and water, while ensuring structural anchoring. With regards to shape and size, bamboo's roots are relatively symmetrical and have diameters ranging from 0.4 to 4.8 mm (Banik 2010). The roots of bamboo are located as a ring around the node areas of the rhizome. The majority of the rooting system can

Fig. 3.5 Rhizomes of bamboo

be observed within 330 mm of the surface of the soil. However, the root of some species can reach depths of 700 mm below the surface (Banik 2015).

3.2.6 Leaves

The bamboo leaf blade differs from culm and branch sheaths by being thinner, more dorsiventral, and possessing petiolate (stalk). The leaf is, however, an extension of the leaf sheath and termed foliage because of its photosynthesis capability. The other leaf type produced by the leaf organ is the cauline leaf. Morphologically, the leaves exhibit parallel venation (Fig. 3.6). Leaf dimorphism is observed in most

Table 3.1 Characteristics of rhizome types

Characteristics	Types of rhizomes	
	Pachymorph	Leptomorph
Appearance	Curved, consists of the thick body and a short neck	Long, cylindrical, and slender
Bud type and placement	Buds appear on both sides of the rhizome. They are relatively large and lateral. Buds may be scaly (formed in summer) or flat. Scaly buds form rhizomes, while flat buds form culms	Buds are lateral in nature. The lateral buds develop into rhizomes or culms
Node and internode	There is no elevation of nodes. Internodes are broad, robust, and asymmetrical	Nodes may or may not elevated or inflated The internode is long, not broad, and often hollow
Rhizome-to-culm ratio	Rhizome thickness greater than that of culm	Rhizome diameter is less than that of culm
Growth direction	Grow horizontal initially and then vertically to form new shoots	
Clump or running	Develops clumps known as sympodial clump	Running
Proximity to domain plant	Close proximity	Branch away from domain plant
Some bamboo species and rhizome type	*Bambusa vulgaris*, *Dendrocalamus strictus*, etc.	Genus *Phyllostachys* and *Pleioblastus*

Source McClure (1966), Gross (2009), Banik (2015)

species of bamboo, and its seasonal changes are caused mainly by the moisture content of the cultivating the soil, the location of leaf on branch, and age (Table 3.2) (Banik 2015).

3.3 Bamboo Culm Anatomy

The anatomical features of the bamboo culm directly influence the properties of the culm. Therefore, understanding bamboo's anatomy is critical to its characterisation. The culm comprises several internodes and nodes whose cell orientations are axial and transverse interconnections, respectively. The nodes have branched vessels capable of radial bending while connecting to diaphragms within the nodes and ensuring a closely knitted culm. The hollow nature of the culm indicates the presence of an inner and outer layer. The primary cell constituent of the inner layer is the sclerenchyma cells, while the outer layer is composed of a two layered epidemics. The outer layered epidermis cells have a waxy (hydroxyl acids), coated,

Fig. 3.6 Distinct leaves of various species of bamboo

cutinized, layered covering, while the inner epidermis is thicker and lignified (Liese 1985a, b; Liese and Tang 2015).

Culms consist of three main components, namely: 50% parenchyma, 40% fibre, and 10% conducting tissues. There are, however, some slight variations dependent on the species

Parenchyma: Ground tissues with vascular bundles incorporated are known as parenchyma cells (Fig. 3.7). Parenchyma cells progressively increase in size and length from the outer culm layer towards the inner layer (Liese and Tang 2015). The parenchyma cells are either thick-walled, long, or vertical (100 × 20 μm) with early shoot lignification or thin-walled, short, cube-like, and unlignified. The retention of cytoplasmic activity resulting from denser cytoplasm of short cube-like parenchyma cells is responsible for the unlignified state (Liese 1985a, b). From culm base to top, the amount of parenchyma decreases as fibre increases.

Table 3.2 Characteristics of bamboo leaves

Species	Seasonal change	Habitat	Location on branch	Age
All		Moist: Larger leaves Dry: Smaller leaves		
Dendrocalamus hamiltonii	March–August: Large leaves Oct.–Nov.: smaller leaves	Moist: Larger leaves Dry: Smaller leaves	Strong branch leaves are bigger than smaller and weaker branches	Large young shoot leaves
Dendrocalamus Asper	March–August: Large leaves Oct.–Nov.: smaller leaves	Moist: Larger leaves Dry: Smaller leaves		
Melocanna baccifera				Larger seedling leaves in comparison to adult culm
Gigantochloa and *amanita*				Larger seedling leaves in comparison to adult culm
Bambusa tulda				Smaller seedling leaves in comparison to adult culm

Adapted from Banik (2015)

The fibres of bamboo are long, slender, forked-end, sclerenchymatous tissue, located in the internode as vascular bundles and sheath around the vessels and, in some species, as solitary strands. Liese (1985a, b) reported a contribution of fibres to tissue mass and weight of 40–50 and 60–70% respectively. A variable length to weight ratio ranging from 150:1 to 250:1 was also observed for both inter- and intra-species. Node and intermodal fibre decrease in length from base to top, however, nodal fibres are shorter than intermodal fibres (Kumar et al. 1994; Tamolang et al. 1980). Physical, mechanical, and chemical properties, such as specific gravity, density, modulus of elasticity, compression, strength, and pulp properties, are strongly influenced by fibre length, density, distribution, and content (Liese and Grosser 1972).

The vascular bundle which is within the culm is critical for water and nutrient transportation (Fig. 3.8). The vascular bundles closer to the outer periphery are

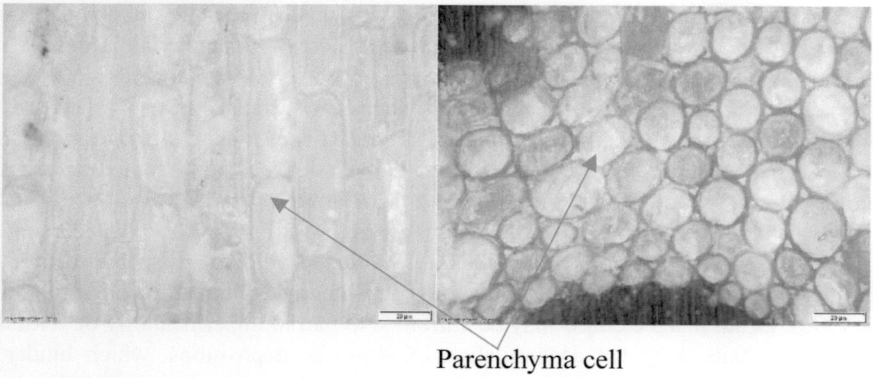

Fig. 3.7 Vertical and cross-sectional views of Parenchyma cells in the culm

small but densely distributed, while those nearer the inner layer are larger but sparsely distributed. From the top of the culm toward the bottom, an increase in the number of vascular bundles is reported, while density decreases simultaneously (Liese 1980a, 1985a, b; Kelemwork 2008). The xylem and phloem mostly make up a vascular bundle.

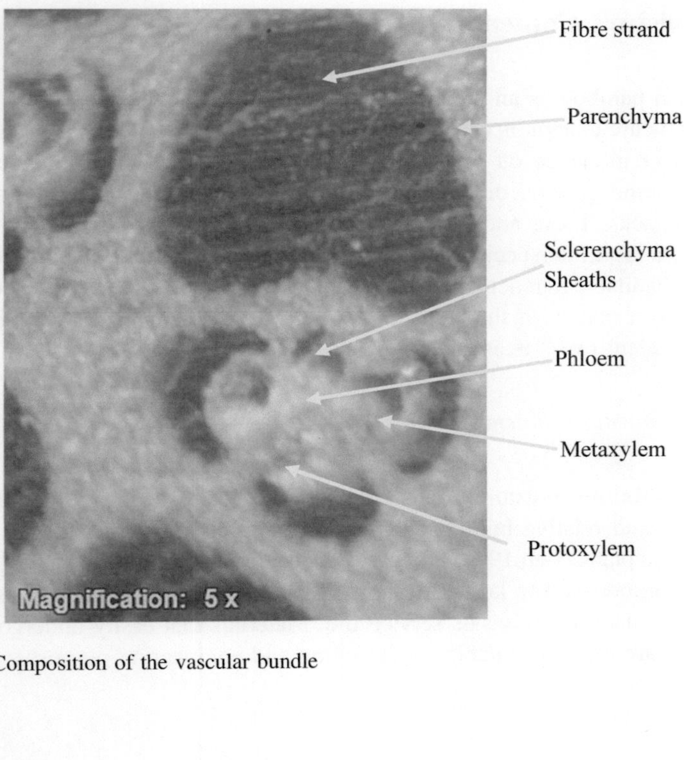

Fig. 3.8 Composition of the vascular bundle

The xylem comprises protoxylem (one or two) which are smaller in comparison to the two larger sized metaxylem vessels separated by parenchyma and axially oriented with varying sizes and length ranging from 40–120 and 200–600 μm respectively (Liese 1985a, b, 1998). The metaxylem's dimensions (area and size) are of critical importance to the survival of the plant. They ensure the transportation of water and treatment and preservation via the sap-displacement method in newly emerging culms. They function throughout the lifetime of the culm.

The phloem is composed of comparatively larger unlignified thin-walled sieve tubes linked to smaller companion cells. A developed phloem is capable of transporting assimilates (food nutrients) for the entire lifetime of the culm (Liese 1985a, b, 1998). Death in aged culms may be attributed to partial impermeability of vessels and sieves caused by the accumulation of gum-like depositions, which hinders conductivity.

A study by Espiloy (1987) showed that vascular bundle properties and features had a significant influence on stress, moisture content, tangent shrinkage, and the modulus of rupture. Thus, it may be concluded that the chemical, mechanical, and physical properties of bamboo are directly influenced by the vascular bundle distribution and density. Sclerenchyma sheaths surround all metaxylem and phloem vessels and tissues. Significant variations in geometry, position along the culm, and species type are observed. (Grosser and Liese 1971, 1973; Wu and Wang 1976; Jiang and Li 1982).

3.4 Moisture Content

To establish bamboo as an engineering material for structural purposes, the influence of moisture content is of vital importance. Moisture content has been found to have a direct influence on the physicomechanical properties of bamboo such as density, specific gravity, dimensional stability, strength, and resistance to fungal and pest attacks. Liese and Grover (1961) concluded that variation in moisture content intra- and interspecies wise, and along the culm, could be attributed to the number of available parenchyma cells. The amount parenchyma cells determine the water storing capacity of the culm. In general, the mathematical expression of for moisture content (MC) is as follows:

$$Moisture\ content\ (MC)\,\% = \frac{wet\ weight - dry\ weight}{dry\ weight} \times 100$$

The equilibrium moisture content (EMC) may be defined as the state where temperature and relative humidity of the air attain equilibrium with the moisture content of the plant (Siau 1995; Forest Products Laboratory 1999). The hygroscopic nature of bamboo makes knowledge of EMC critical, as bamboo will experience moisture fluctuations during its service life. Materials that easily undergo changes in moisture are often unsuitable for structural and engineering purposes.

Table 3.3 Moisture contents

Source	Bamboo species	Findings–Results
Liese (1985a, b), Latif and Liese (2002a, b), Islam et al. (2002), Kamruzzaman et al. (2008)	*Dendrocalamus strictus*	The MC of one-year-old shoots = 120–130% for the base and top part For older culms (3–4 years), MC is higher at the bottom (100%) than at top (60%) Higher MC in the inner section of culm wall than the outer layer
Sulthoni (1989), Sattar et al. (1994), Hamdan et al. (2007), De Vos (2010)		The equilibrium moisture content (EMC) of bamboo is found to be comparable to that of wood. This implies that wood EMC data can be adapted for use in bamboos
Qisheng et al. (2001)	*Prosopis pubescens*	MC is dependent on seasons and species type Cutting age MC = 80% MC after air seasoning = 15.7
Razak et al. (2007)	Four-year-old *Bambusa vulgaris*	MC for green bamboo = 89%
Hamdan et al. (2009)	*Gigantochloa scortechinii*	At harvesting age, the average MC = 90%
Wakchaure and Kute (2012)	Three-year-old *Dendrocalamus strictus*	Reported that all physicomechanical properties of bamboo for structural purposes are directly affected by MC
		MC decreases along culm from base to top for both green and seasoned bamboo Compressive strength increased with decreasing MC: (16.09%: 54.29 MPa) moreover, (9.98%: MPa)

For moisture content determination, the ASTM D 4933-39 (ASTM 1997) may be used as a guide. A summary of some research conducted on the moisture content is collected in Table 3.3.

3.5 Specific Gravity (Density) and Mechanical Properties

The use of bamboo for load-bearing structural members is gaining prominence in many developing countries. Studies have reported that bamboo has outstanding mechanical properties, in some instances comparable to steel. These properties, however, have a direct correlation with specific gravity, which is also dependent on the fibre content within the culm (Table 3.4). Liese (1985a, b) reports that the specific gravity of bamboo shows a variation of 0.5–0.8 (0.9) and is higher within the outer layer of culm wall than the inner layer. Specific gravity is found to increase from base of culm to the top, and is higher in nodes than internodes. The high specific gravity (density) and mechanical strength within the outer layer, top part, and nodes of culms are due to lower parenchyma but higher fibre located in these sections (Janssen 1981a, b; Liese 1985a, b).

Table 3.4 The influence of density on the physical and mechanical properties and the anatomy of bamboo

Mechanical properties	Influence of physical properties	Anatomy/Morphology influence
General mechanical properties	Mechanical properties increase with decreasing moisture content As density increases, mechanical properties also increase	With the exception of modulus of elastic, every other strength property positively correlates with vascular bundle distribution An increase in sclerenchyma and conductive cells leads to an increase of density Fibre diameter correlates very positively with mechanical properties
Compression Bending stress Elastic modulus		The size of vascular bundles and fibre length correlate positively with the stated mechanical properties. In vascular bundles, as tangential size increases, strength properties decrease
Shear strength Maximum strength		Shear strength decreases with increasing fibre length. Cell-wall thickness or density are the attributing factors, not the parenchyma-fibre percentage The maturity and increase in wall thickness results in a hardened culm ensuring maximum strength
Compression Bending stress Elastic modulus Modulus of rigidity		Only modulus of rigidity correlates negatively with cell-wall thickness; all other stated properties correlate positively
Stiffness	As specific gravity increases, so do strength and stiffness	
General strength of bamboo		Strength increases with increased age. The strength increment is due to increase of matured tissues

Source Janssen (1981a, b), Rangqui and Kuihong (1987), Razak et al. (1995), Gnanaharan et al. (1994), Hamdan (2004)

3.6 Chemical Composition

The application of bamboo for biofuels, such as firewood, charcoal, pellets, ethanol, and methane, as possible alternatives to traditional fossil fuels, demands a thorough analysis of its chemical composition. The constituents of bamboo may be categorised as main or minor. The main chemical constituents are cellulose, hemicellulose, and lignin, constituting 90% of the overall mass of the culm. The cellulose is composed of β-1-4-linked glucose anhydride units characterised by linear chains. Cellulose content in bamboo is about 40–60% (Liese and Tang 2015). Xylan is built from β-1-4 linked xylose units having 4-0-methyl-D-glucuronic acid, L-arabinose, and acetyl groups as side chains (Puls 1992). It constitutes 90% of hemicellulose and accounts for a quarter of the culm-cell material (Vena et al. 2013; Liese and Tang 2015). Lignin is the second most prevalent chemical compound (20–26%) in bamboo and is a three-dimensional polymer macromolecule (Li 2004; Liese and Tang 2015).

The minor constituents are namely >a small quantity of ash, resins, several polysaccharides, proteins, tannins, inorganic salts, and waxes (Li 2004; Liese and Tang 2015). Vena et al. (2013) observed that the chemical composition is similar to other woods, however (Tomolang et al. 1980; Chen and Qin 1985) reported that bamboo had higher levels of ash content, silica content, and alkaline extractives in comparison to wood. The constituents vary with respect to age, growth environment, and species type. The concentration of lignin and carbohydrate vary with age (Liese and Tang 2015). Yusoff et al. (1992) reported that the holocellulose content did not significantly vary with age for *Gigantochloa scortechinii* (Aged 1–3 years), however alpha-cellulose, ash and silica content, extractives, pentosan, and lignin increase with bamboo age. Other constituents are deoxidized saccharide (2%), fat (2–4%), protein (0.8–6%), and starch (2–6%) (Li 2004).

The percentage of starch is very critical to the culm's susceptibility to fungal, mould, and borer attacks. Durability is, therefore, a function of the starch content (Liese 1980a, b). Maximum starch levels are realised in the dry seasons but lower in the wet seasons. Starch levels also depend on the species. Silica content is mostly concentrated in the epidermis and increases from the base (1%) to the top (6%). Pulp properties are influenced by the silica content (Liese and Latif 2000). A higher concentration of ash content (1–5%) is distributed within the inner layer of the culm than the outer. The main constituents of bamboo ash content are calcium, inorganic minerals, potassium, and silica. Table 3.5 summarises the chemical composition of some bamboo species of various lengths, as well as the composition of their culms.

Table 3.5 The chemical composition analysis of some bamboo species

Species	Chemical composition (%)					
	Holocellulose	Lignin	Ash	Alcohol-benzene solubility	Hot-water solubility	1% NaOH solubility
Bamboo blumeana[a]	81.6	27.2	4.7	7.5	8.2	28.6
Bambusa blumeana[e]	69.2	21.6	1.3	3.8	7.3	23.3
Bamboo spp[a]	79.4	25.3	1.9	4.0	11.4	25.0
Dendrocalamus asper[c]	74.0	28.5	1.5	5.5	9.2	24.7
Dendrocalamus strictus[a]	78.3	27.7	2.0	6.7	12.4	26.6
Gigantochloa albociliata[a]	80.0	26.0	1.8	4.8	9.9	25.7
Gigantochloa scortechinii[d]	67.4	26.4	1.3	3.4	5.9	19.4
Phyllostachys pubescens[b]	71.7	23.6	1.4	4.6	9.2	24.7
Schizostachyum zollingeri[e]	71.6	21.4	1.3	2.5	5.1	24.3

Sources
[a]Ratanophat (2004)
[b]Li et al. (2007)
[c]Kamthai (2003)
[d]Jamaludin et al. (1992)
[e]Nor Aziha and Azmy (1991)

3.7 Mechanical Characterisation

Bamboo is unique among timbers because its physical and mechanical properties vary significantly with moisture content, age, diameter, type, position along the culm, and length or height of the culm. Literature and various research indicate that from two-and-a-half to four years the optimal strength of bamboo is attained, beyond which strength begins to decrease (Amada and Untao 2001; Lo et al. 2004). Studies by Amada et al. (1997) suggest that bamboo fibres have a tensile strength equivalent to that of steel. It must, however, be noted that this finding contradicts other studies. As a reinforcement in matrices such as concrete, Ghavami (1995) observed an ultimate strength four times that of concrete without reinforcement. Several mechanical testing procedures are available. However, tensile, compression, fatigue, impact shear torsion, hardness, and the wear tests are described in this section.

3.7.1 Standardisation

The purpose of standardisation of test procedures, such as material testing methods, is to determine the accuracy of design parameters or values and establish a reference framework common to research in the community. Standardisation, therefore, seeks to satisfy both technical and social objectives of test procedures. Acceptance of standardised testing procedures is attained through a compilation of comparable test data for better understanding and refinement through technical statistical analysis to enhance confidence in design results and sustain advocacy to ensure social acceptance.

3.7.2 History of Bamboo Standardization

Standardisation of mechanical properties for bamboo was developed in 2004 as a collaborative effort between the International Organization for Standardization (ISO) and the International Network for Bamboo and Rattan (INBAR). The standards were classified as follows:

- ISO 2004a: structural design standards of bamboo
- ISO 2004b and ISO 2004c: for determination of the mechanical properties of bamboo.

The four main ISO standard approved tests include, tensile strength parallel to grain, the compression strength of a full culm, longitudinal shear, and the three-point flexural tests. Harries et al. (2012) tabulated the standards which are simplified and adapted in Tables 3.6 and 3.7.

3.7.3 Tensile Testing

Tensile testing can also be referred to as tension test and is one of the most basic fully standardised and inexpensive mechanical tests available. In tensile testing, parallel to grain of bamboos, the following standards can be used, namely: ASTM 143-94, ASTM 3500-90, and ISO 3345. Early researchers such as Janssen (1981a, b) and Arce-Villalobos (1993) observed varying tensile-strength results from several sources of literature ranging 100–400 N/mm^2 as shown in Table 3.8.

Janssen (1981a, b) and Arce-Villalobos (1993) observed very high tensile values while the modulus of elasticity was very low with accompanying low shear stress. These limitations of tensile testing, especially the low modulus of elasticity, may be attributed partly to induced failure from the grips due to longitudinal shear in the inter-fibre plane or transverse compression. The recommended tensile-testing specifications proposed the use of a dog-bone shaped specimen or a reduced gauge

Table 3.6 Current standards for bamboo testing

Test	Equation and variables	Notes
ISO 2004b: Tension parallel to culm	$\sigma_t = \frac{F_t}{A_t}$ $\sigma_t = tensile\ stress\ parallel\ to\ fibre$ $F_t = tensile\ loading$ $A_t = gauge\ area$	The dog-bone shape is utilised for specimens. Local damage due to gripping must be avoided
ISO 2004b: Compression parallel to culm	$\sigma_c = \frac{F_c}{A_{culm}}$ $A_{culm} = \frac{\pi}{4}\left[d^2 - (d-2t)^2\right]$ $\sigma_c = Ultimate\ compressive\ stress$ $F_c = compressive\ loading$ $A_{culm} = net\ area\ of\ culm$ $d = outside\ diameter$ $t = specimen\ thickness$ $l = culm\ length,\ l \leq 2d$	20–80% of the stress-strain curve is used in calculating the modulus due to compression, while the average strain is preferred. Friction between the culm and loading head must be minimal
ISO 2004b: Shear parallel to culm	$\tau_z = \frac{F_z}{4lt}$ $\tau_z = shear\ strength\ parallel\ to\ fibre$ $F_z = shear\ force$ $l = specimen\ length,\ l = d$ $d = outer\ diameter\ of\ culm$ $t = specimen\ thickness$	A butterfly- or bowtie-shaped sample is used Specimens with or without nodes are tested separately. Caution must be used to ensure culm ends are smooth, parallel, and right-angled to the longitudinal axis of the culm. Uniformity in distributed load is ensured by the use of rigid loading blocks

Adapted from Harries et al. (2012)

length, such as in ISO 3345. Arce-Villalobos (1993) mathematically proved that the standard dog-bone shape might not be the best for bamboo but rather a straight specimen may be preferred.

The effects of tensile force and grip influence on the transition zone and gauge region of a dog-bone shaped specimen are shown in Fig. 3.9, where two elemental strips are used in the analysis. It can be deduced that the increase in the x-direction for both sides of $2x = W + 2Cy$, where W = *width of gauge section*. The increase on one side in the x-direction (Fig. 3.10) can be expressed as

$$x = \frac{W}{2} + Cy \tag{3.1}$$

As a tensile force P_y is applied, the specimen will deform in the y-direction. Assuming an ideal proportional characteristic behaviour, then the elemental strips will deform as follows:

$$\left(\varepsilon_y\right)_1 = \frac{\frac{P_y}{Wt}}{E_y} \tag{3.2}$$

Table 3.7 Other recognised tests

Test	Equation and variables	Notes
ISO 2004b: Bending perpendicular to fibre	$f_b = \frac{F_b l d}{12 I_b}$ $E_b = \frac{l^3}{56 I_b \Delta}$ $I_b = \frac{\pi}{64}\left[d^4 - (d-2t)^4\right]$ f_b = *apparent modulus of rapture* F_b = *bending force* l = *length of culm*, $l \geq 30d$ d = *outer diameter* I_b = *moment of inertia of culm* E_b = *bending modulus of elasticity* Δ = (*measured mid-span deflec.*) t = *specimen thickness*	All calculated results are termed as apparent or effective; however, they can be used in bamboo culm designs. The three-point bending or flexural test is used
Longitudinal shear flow (Harries et al. 2012)	$\tau = \frac{F_b\left[\left[d^4-(d-2t)^4\right]\right]}{48 I_b t}$ τ = *shear stress* F_b = *bending stress* d = *culm diameter* t = *specimen thickness* I_b = *moment of inertia* *notch depth* = 0.5 d	The test is a variant of the bending-perpendicular-to-fibre test, however with a perpendicular notch placed at the end of the constant moment region on the tension side. The notch is expected to initiate splitting failure longitudinal to the axis of the culm
Interlaminar shear (Moreira 1991; INBAR 1999)	$\tau_{int} = \frac{F_{int}}{A_{int}}$ τ_{int} = *interlaminar shear stress* F_{int} = *interlaminar shear force* A_{int} = *shear plane area*	An s-type specimen shape is adopted. During tension, pure shear at the centre of the specimen is induced. Local stresses caused by grip forces should be avoided
Shear perpendicular to fibre (Cruz 2002)	$\tau_{\perp} = \frac{F_{\perp}}{A_{\perp}}$ $\tau_{\perp}$ = *Transverse shear* $F_{\perp}$ = *shear force* $A_{\perp}$ = *specimen area*	A simple two-shear arrangement is utilised. Specimen failure results in three broken pieces
Pin shear (Janssen 1981a, b; Sharma 2010)	*Shear* ($\theta < \sim 30°$): $\tau_b = \frac{F_b}{4bt}$ *Bearing* ($\theta < \sim 60°$): $\sigma_b = \frac{F_b}{2d_b t}$ τ_b = *shear stress* F_b = *shear force* b = *gap* t = *thickness of specimen* σ_b = *stress on bearing* d_b = *outer diameter of culm*	A simple bolted culm configuration is used. Block shear or bearing failure may occur
Split-Pin transverse tension (Mitch et al. 2010)	$\sigma_{\perp} = \frac{P}{[2t(2w-2a)]}$ $K_I = \frac{P\sqrt{\pi a}}{4wt}\sqrt{\frac{2w}{\pi a}tan\left(\frac{\pi a}{2w}\right)}$ $\sigma_{\perp}$ = *transverse tension* P = *applied load to cause failure* t = *culm wall thickness* $2w$ = *specimen length* $2a$ = *length of equivalent crack*	The notches and holes within the specimen represent crack initiation

(continued)

Table 3.7 (continued)

Test	Equation and variables	Notes
	$K_I = Mode\ 1\ crack\ intensity\ factor$	
Edge Bearing (Amada et al. 1996; Torres et al. 2007; Mitch et al. 2010)	$f_{rNS} = \frac{M_{NS}(c+h)}{I}$ $f_{rEW} = M_{NS}\frac{(c+h)}{I} - \frac{pl}{2lt}$ $M_{NS} = \frac{pld}{2\pi}\left(1 - \frac{4}{AD^2}\right)$ $M_{EW} = M_{NS} - \frac{pld}{4}$ $c = \frac{t}{2},\ I = \frac{lt^3}{12}$ $h = \frac{d}{2} - \frac{2c}{ln\left(\frac{d}{2c} + \frac{1}{\frac{d}{2c} - 1}\right)}$ $E_\varphi = \frac{pld^3}{8I\Delta}\left(\frac{\pi k_1}{4} - \frac{2k_2^2}{\pi}\right)$ $where\ k_1 = \approx 1 + \frac{4.24I}{ltd}$ $and\quad k_2 = 1 - \frac{4I}{ltd^2}$ $f_{rNS} = Transverse\ modulus\ of\ rapture$ $p = applied\ load\ (N/m)$ $l = length\ of\ specimen$ $t = culm\ wall\ thickness$ $E_\varphi = transverse\ modulus\ of\ elasticity$ $\Delta = vertical\ deflection$	This testing method can be used as a substitute for the split-pin test. During testing, a full culm is subjected to compressive loading along the longitudinal axis of the culm. This test is simple to perform and requires little sample preparation and is therefore well suited to field testing difficult-to-access transverse materials

Adapted from Harries et al. (2012)

Derived from:

$$E_y = \frac{\sigma_t}{\varepsilon_y} \tag{3.3}$$

$$\sigma_t = \frac{P_y}{A} \tag{3.4}$$

$$A = Wt \tag{3.5}$$

where

A = *Cross-sectional area of gauge section*
E_x = *Modulus of elasticity in the x-direction*
E_y = *Modulus of elasticity in the y-direction*
ε_x = *strain in the x-direction*
ε_y = *strain in the y-direction*
σ_t = *tensile stress*
τ = *shear stress*
γ = *Shear deformation angle [radians]*
P_y = *force due to tension in the y-direction*
t = *thickness of specimen*
v = *poisson's ratio*
W = *width of specimen.*

Table 3.8 Early fundamental literature on tensile strength of bamboo

Author	Year of pub.	Method and tensile strength or findings
H.F. Meyer and B. Ekelund	1923	**Method**: Tested bamboo-reinforced beams using steel-reinforced concrete-beam formulae **Results**: 100 N/mm^2
C.H. Duff	1941	**Method**: Strips from *Phyllostachys pubescens* culms were tested **Results/findings**: Outer layer = 342 N/mm^2 Inner layer = 54 N/mm^2 Strength at Node = 80% strength at internode
M. Ota	1953	**Method**: Studied the effects of moisture content on the tensile strength at failure for *Phyllostachys pubescens* and *Phyllostachys edulis* **Findings**: A linear relationship between moisture content and tensile strength
K.P. Karamchandani	1959	**Method**: To explore bamboo's suitability as a construction material, *Dendrocalamus strictus* was used **Results**: Outer layer: 100–335 N/mm^2 Inner layer: 150–160 N/mm^2
Cox and Geymeyer	1969	**Method**: The study focused on the tensile strength of bamboo's (*Arundinaria tecta*) relationship to parameters such as age, soil type, morphology, and external influences **Results**: Modulus of elasticity = 18,670 N/mm^2 Max Tensile stress = 110 N/mm^2
J.L. Atrops	1969	**Method**: The study used a dog-bone shaped specimen **Results**: Outer layer: 290 N/mm^2 Inner Layer: 153 N/mm^2
E.C. McLaughlin	1979	**Method**: *Bambusa vulgaris* was used to study the relationship between tensile strength and density **Findings**: Proved that the tensile strength of bamboo directly correlated with density
L. Xiu-Xin and W. Liu-Keqing	1985	**Method**: The bamboo species *Phyllostachys* harvested from four habitats in China were used to study the mechanical properties in relation to tensile strength, age, and density **Findings**: Regression equations relating tensile strength to age were developed
T. Soeprayitno et al.	1988	**Method**: Experimentation was carried out on *Gigantochloa pseudoarundinacea* from hilly and valley locations **Result**: Hill slope: a. Tensile strength: 177.9 N/mm^2 b. E-modulus: 27,631 N/mm^2 Valley: Tensile strength: 149.4 N/mm^2 a. E-modulus: 19,643 N/mm^2

(continued)

Table 3.8 (continued)

Author	Year of pub.	Method and tensile strength or findings
S. Prawirohatmodjo	1988	**Method**: Determination of the tensile strength of Indonesian bamboos (*Bambusa vulgaris, Dendrocalamus, Gigantochloa apus, Gigantochloa atter,* and *Gigantochloa verticilliata*) **Results**: Green bamboo: 297 N/mm^2 Dry bamboo: 315 N/mm^2
A. Sharma	1990	**Method**: This study explored the possible use of bamboo for concrete reinforcement in T-beams. *Bambusa vulgaris* was used for this study **Results**: Parallel tensile strength (Node): 145 N/mm^2 Parallel tensile strength (Internode): 200 N/mm^2

Source Janssen (1981a, b) and Arce-Villalobos (1993)

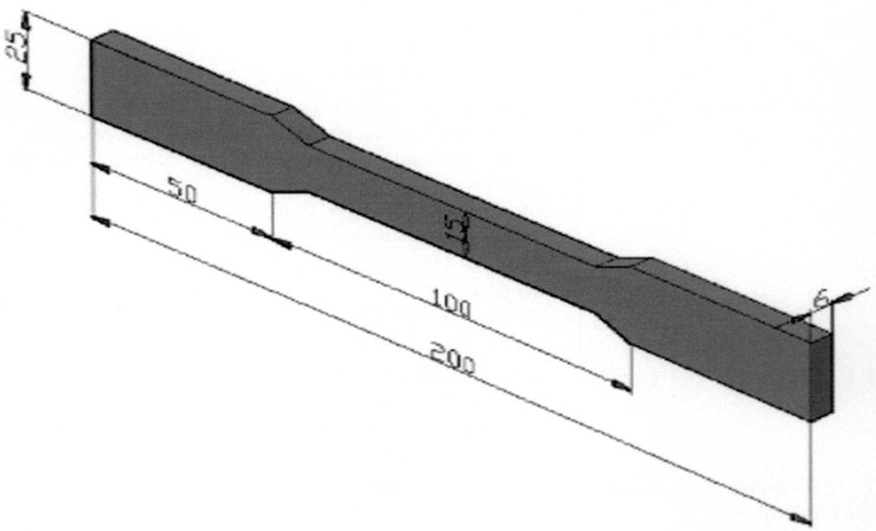

Fig. 3.9 A dog-bone-shaped specimen

$$\left(\varepsilon_y\right)_2 = \frac{\frac{P_y}{(W+2Cy)t}}{E_y} \tag{3.6}$$

Because of the Poisson's effect strips, strips from Eqs. 3.1 and 3.2 will deform in the lateral or x-direction into:

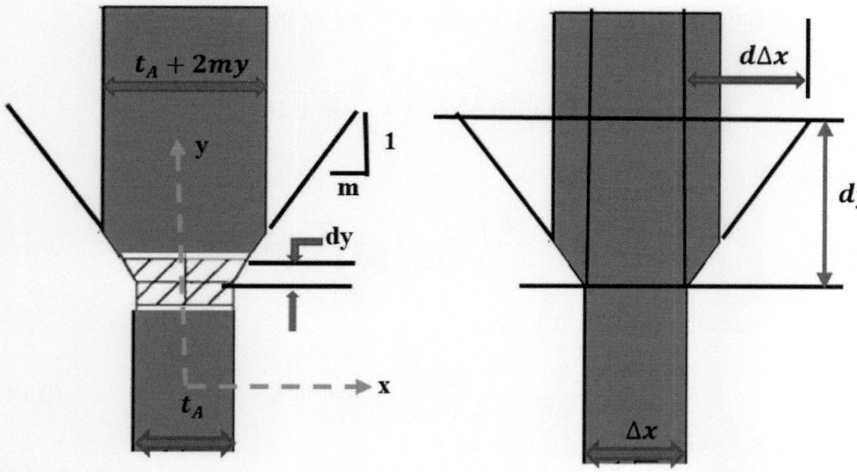

Fig. 3.10 Section of the specimen

$$(\varepsilon_x)_1 = \nu \frac{\frac{P_y}{Wt}}{E_y} \tag{3.7}$$

and

$$(\varepsilon_x)_1 = \nu \frac{\frac{P_y}{(W+2Cy)t}}{E_y} \tag{3.8}$$

An appearance of shear deformation is observed when the difference in lateral deformation in the elementary strips defined in Eqs. 3.1 and 3.2 are considered. The shear angle deduced from Fig. 3.9, produced as a result of the difference in deformation, is given as:

$$d\gamma = \frac{\frac{d\Delta\epsilon_x}{2}}{dy} \tag{3.9}$$

The shear stress is derived as:

$$d_\tau = \frac{E_x}{2(1+\nu)} d\gamma \tag{3.10}$$

Substituting Eqs. (3.7) and (3.8) into Eq. (3.9) gives:

$$\frac{\Delta\varepsilon}{2} = \frac{P_y \nu}{2tE_y}\left(\frac{1}{W} - \frac{1}{W+2Cy}\right) \tag{3.11}$$

Substituting Eq. (3.11) into Eq. (3.10) turns out to be

$$d_{\tau} = \frac{E_x}{2(1+v)}\frac{d}{dy}\left[\frac{P_y v}{2tE_y}\left(\frac{1}{W} - \frac{1}{W+2Cy}\right)\right] \tag{3.12}$$

Integrating Eq. (3.12) gives:

$$\tau = \sim \frac{E_x P_y v C}{2(1+v)tE_y}\int \frac{dy}{(W+2Cy)^2} \tag{3.13}$$

Then integrating Eq. (3.13) yields:

$$\tau = \frac{E_x P_y v}{4(1+v)(W+2Cy)E_y t} + \tau_i \tag{3.14}$$

The assumption can be made that no change in shear stress occurs for regions below line A–A, so:

$$\tau_i = 0 \tag{3.15}$$

However, the shear stress value for the root of the gauge region is obtained by finding the limit of Eq. (3.14):

$$lim_{y-0}\tau = \frac{E_x v N_y}{4(1+v)E_y t_A} \tag{3.16}$$

$$\tau = \frac{E_x v \sigma_t}{4(1+v)E_y} \tag{3.17}$$

Inputting data into Eq. (3.17) gives values that are very close to the maximum stress of bamboo and, hence, using the dog-bone shape might not be the best. However, many researchers have solved this problem by introducing bonded tabs and using bonded tabs to reduce the effects of the grips.

Sharma et al. (2015) compared the tensile strength parallel to the grain of several softwoods with bamboo, and the results are presented graphically in Fig. 3.11.

3.7.3.1 Modulus of Elasticity

The modulus of elasticity is determined by measuring the material's stiffness. It, however, is limited to only the linear section of the stress-strain curve. The specimen material is expected to return to its original state when the tensile load applied to the linear section is removed. Hook's law only applies to this region of the curve. However, the ultimate tensile strength is measured by determining the maximum

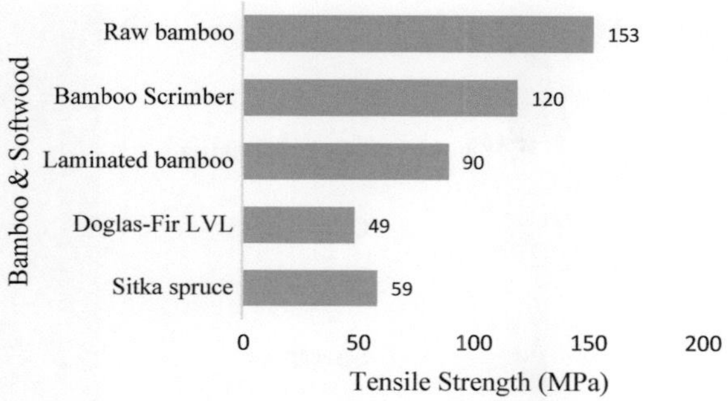

Fig. 3.11 Tensile strength parallel to grain of several softwoods and bamboos Data from Sharma et al. (2015)

load that a specimen is capable of resisting during tensile testing (Muller and Rebelo 2015).

3.7.3.2 Experimental Work

A dog-bone specimen configuration was used in conducting the test in compliance with ISO 3345. The Instron Universal testing machine of capacity 100 kN and a loading rate of 1 mm/min was used (Fig. 3.13). The specimen dimensions are shown in Figs. 3.12.

3.7.3.3 Bamboo Culm Morphology: Results and Discussion

Figure 3.14 shows the stress-strain curve of *Bambusa arundinacea* with nodes. The ultimate tensile strength was 114 MPa with an elastic modulus of 126.2 MPa. The stiffness of the bamboo is determined by the elastic modulus and derived from the linear region of the stress-strain curve. Hook's law as shown in Eq. 3.18 can, therefore, be applied to this region.

$$Elastic\,modulus\,(E) = \frac{Stress\,(\sigma)}{Strain\,(\varepsilon)} \quad (3.18)$$

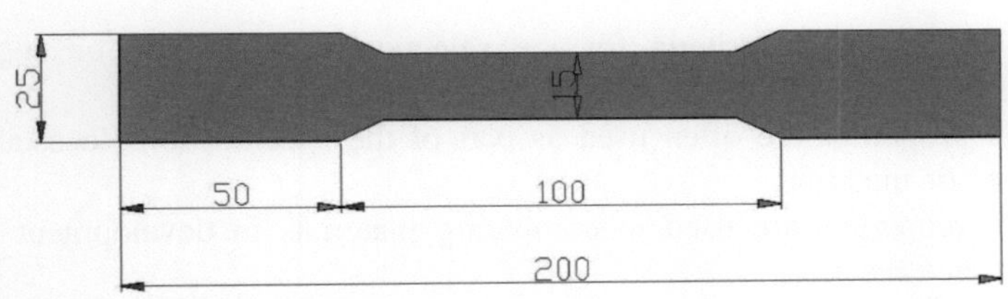

Fig. 3.12 Dog-bone-shaped specimen

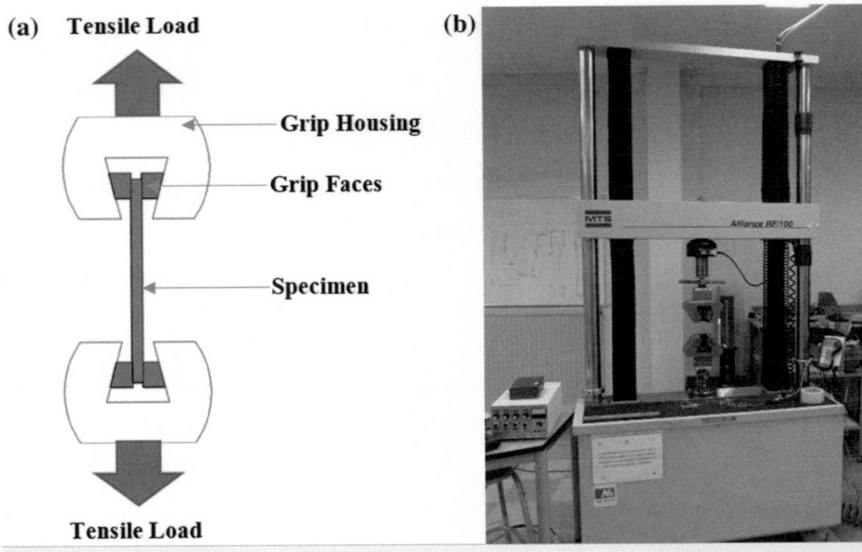

Fig. 3.13 **a** Schematic of wedge grips self-tighten. **b** Tensile equipment

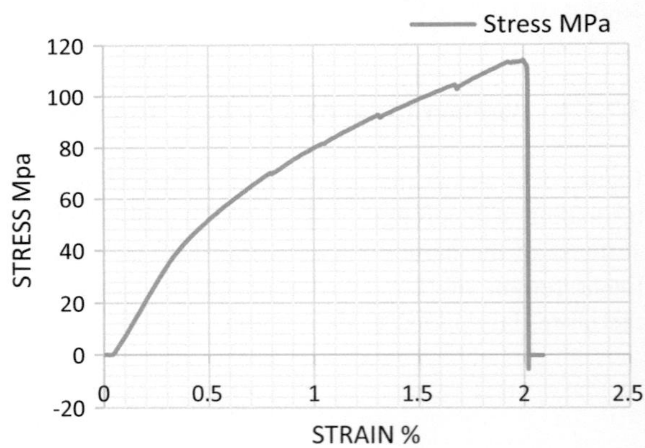

Fig. 3.14 Stress-strain curve of *Bambusa arundinacea* with nodes

The reasons for tensile testing are as follows:

- Used as parametric criteria for the selection of materials for engineering purposes.
- Tensile properties are often used as part of the specifications to establish the qualities of materials.
- Tensile properties are used in comparing materials in development to those already in existence.

Tensile properties can be used as predictive tools in determining the behaviour of materials subjected to loading.

3.7.4 Compressive Testing

Extensive and fundamental works have been conducted in determining the compressive strength of bamboo. A thorough review of these studies has been reported by Janssen (1981a, b) and Arce-Villalobos (1993). A summary is presented in Table 3.9.

Sharma et al. (2015) compared the tensile strength parallel to the grain of several softwoods with bamboo and tabulated the results. See Fig. 3.15 for a graphical representation of the results.

3.7.4.1 Experimental Works

Compression tests involve subjecting the samples to crushing loads. The ultimate compressive strength is calculated, as the elastic limit is derived from the compressive stress-strain curve. A compression test parallel to the grain (to the axis of the fibres) is conducted on bamboo culm as shown in Fig. 3.16. The culm specimen was machined in accordance with ISO 3787 on the compressive machine as illustrated in Fig. 3.16 with a hemispherical head and applying a gradual, continuous load at a constant rate of 0.01 mm/s.

3.7.4.2 Results and Discussion

From experimentation, the ultimate stress in compression parallel to the grain for the species *Bambusa vulgaris* was 113.54 (specimen with node) and 98.10 (specimen without node) as observed in Fig. 3.17 and Table 3.10.

3.7.5 Fatigue Testing

Fatigue testing is defined as the application of a repeated cyclic load to a test specimen. The application of the cyclic load may be in three forms, namely: in-service load simulation and fixed or repeated loads. The load is repeated a million times or many hundreds of times to obtain a realistic and accurate result. The main purpose of this testing is to establish or predict the fatigue life of the material, which in this case is bamboo. A graph of stress amplitude to failure cycles is plotted to show the crack initiation, propagation and the eventual point of failure.

Table 3.9 Early fundamental literature on compressive strength of bamboo

Author	Year of pub.	Method and compressive strength (σ_c) or findings
H.F. Meyer and B. Ekelund	1923	**Method**: Lead cushion or direct contact between test equipment were used. Failure was described as a series of vertical cracks **Results**: σ_c = 44 and 39.27 N/mm^2
Espinosa	1930	**Method**: *Bambusa Spinosa* specimens of height 1200 and 350 mm were used **Result**: 1200 mm specimen: 27–32 N/mm^2 350 mm specimen: 51.8–82.8 N/mm^2
M. Ota	1950	**Method**: Small cube specimens obtained from the inner, outer, and complete wall thickness of *Phyllostachys edulis* and *Phyllostachys reticulata* with moisture contents between 10 and 14.7% were used **Results/findings**: Compressive strength decreases as the node is approached. Species of relative densities 965 and 866 kg/m^3 had compressive strengths of 81.6 and 83.1 N/mm^2, respectively
Limaye	1952	**Method**: *Dendrocalamus strictus* was used for this study to determine the influence of moisture content, age, and specimen position on bamboo **Findings**: A higher modulus of elasticity at the bottom position compared to the top and middle positions of the bamboo was observed. Seasoning was found to increase the compressive strength by 40% and was therefore a significant factor. Strength also increased with age, up to the third year. A compressive strength of 44 N/mm^2 was obtained
Sekhar and Rawat	1957	**Method**: First, attempt at standardising the test process by using a 10, or lower, length-to-wall thickness ratio. The study recommended the use of two samples with one kiln dried to 12% and the other green (both obtained from the internode). *Bambusa nutans* was used in the study
J.L. Atrops	1969	**Method**: Three specimen options with internodes, node positioned at centre (**Nc**), and node at extreme ends (**Ne**) were used. The moisture content of all specimens averaged 18.1% and specimen size selected according to a 1:4 diameter to height ratio **Results**: Min and max compressive strengths were obtained: Internode: 402 and 522 N/mm^2 **Nc**: 407 and 537 N/mm^2 **Ne**: 528 and 433 N/mm^2
Janssen	1981a, b	**Method**: The study included several series of tests on *Bambusa blumeana* to determine the compression capacity with an emphasis on moisture content (MC), node and internode, and position along culm **Results**: σ_c = 60–176 N/mm^2

(continued)

Table 3.9 (continued)

Author	Year of pub.	Method and compressive strength (σ_c) or findings
		For green bamboo: $\sigma_c = 0.0075\rho$ For 12% MC: $\sigma_c = 0.094\rho$ where ρ = density
L. Xiu-Xin and W. Liu-Keqing	1985	**Findings**: An equation relating compressive strength to age at harvesting was derived: $\sigma_c = 588.4 + 53.9t - 4.75t^2$ where t = age of culm at the time of harvesting
Sotela	1992	**Method**: *Guadua* sp. was used in this study where bamboo was compared with some hardwoods. Bending test was employed in the determination of the elastic modulus **Result**: 41.1 N/mm^2

Source Janssen (1981a, b) and Arce-Villalobos (1993)

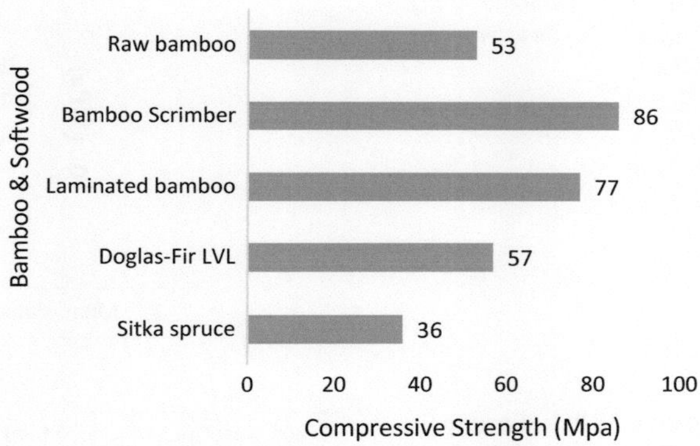

Fig. 3.15 Comparison of compression parallel to the grain. *Data source* Sharma et al. (2015)

$$fatigue\ life = \frac{Total\ number\ of\ cycles}{failure\ under\ specified\ laoding\ conditions} \quad (3.19)$$

3.7.6 Impact Testing

The current ethics of engineering efficiency requires cost reduction in manufacturing while increasing the quality output and ensuring that safety, liability, and reliability standards are maintained. Impact testing is one key experiment tool required for quantifying impact resistance. Impact testing is one of the most

Fig. 3.16 Shows sample of bamboo culms and a compressive test machine with a sample

Fig. 3.17 Ultimate strength (σ_C) of *Bambusa vulgaris* (BV)

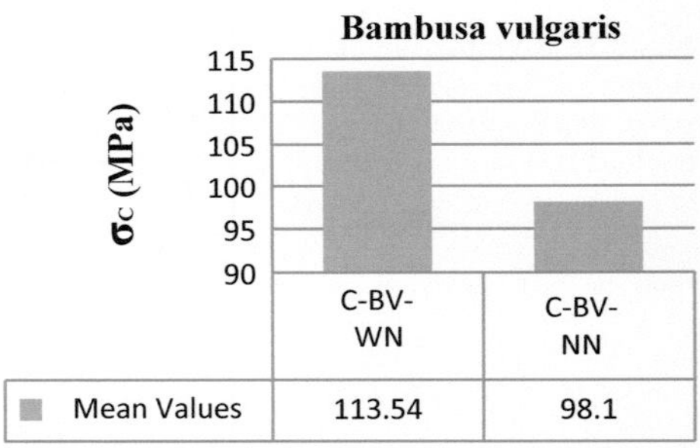

Table 3.10 Ultimate strength (σ_C) of *Bambusa vulgaris* (BV)

Node description	Mean	Standard deviation
With Node (WN)	113.54 MPa	0.25
No Node (NN)	98.80 MPa	1.29
Increase in σ_C	15.74%	

important tests for engineering materials that operate in environments that subject them to possible impact from high-velocity projectiles and objects. During testing, the bamboo is subjected to very high continuous loads, and its resistance to the impact is measured by determining the amount of energy absorbed (material strength and ductility) while fracturing occurs at relatively high velocities. Impact testing is vital in assessing the service life of most parts. A load versus displacement curve is constructed to determine the energy absorbed, which is seen in the area under the curve. A schematic representation of the Charpy impact test is shown in Fig. 3.18.

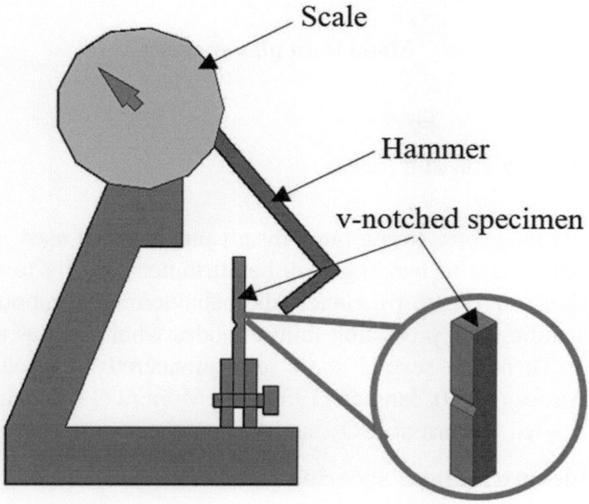

Fig. 3.18 Charpy impact resistance test

Impact testing helps in determining the following:

1. The impact energies that specimen or structural part will experience during its service lifetime
2. The specific impact type required for the delivery of that energy
3. The selection of material types capable of resisting the impact over the life-span projection.

Impact properties are affected by the following:

1. Stresses from moulding process
2. Orientation of polymer
3. Weak regions within material or part (weld lines, etc.)
4. Part geometry
5. Addition of additives (colour agents, etc.).

3.7.7 Shear Testing

Shear

Bamboo has been used in the arts and craft for items such as decorative pieces, furniture, baskets, and woven mats for many centuries, because of the ease of splitting. The ease with which bamboo splits can be attributed directly to its shearing behaviour. Early researchers such as Meyer and Ekelund (1922) made the observation, that in comparison other woods, bamboo has comparable bending and tensile strength; however, bamboo possess weaker shear strength of about 8% of its compression strength. Although in handicrafts bamboo's weak shearing characteristic is

Fig. 3.19 Model II fracture

advantageous, it is a huge disadvantage when used as a structural part. In reality, failure due to bending may be attributed directly to shear between bamboo fibres (Janssen 1981a, b). Hence, the behaviour of bamboo in shear is critical to understanding and predicting failure modes when used as a structural member.

There are several shear tests pioneered by researchers including Ota (1950), Atrops (1969), Janssen (1981a, b), Moreira (1991), and Cruz (2002). Their research has led to current ISO standards on shear tests for bamboo.

Flexure induced shearing

Bamboo under flexure fails mainly by splitting or longitudinal shear. This is because the orientation of bamboo fibres are longitudinal in direction. This splitting induced by flexure is very problematic. However, a weak lignin matrix, the absence of tangential fibres, and high shrinkage rate from drying can be other sources of splitting in bamboo (Richards 2013). Flexure-induced shearing is the main failure mode, consisting of a combination of both mode I (tangential tension) and mode II (in-plane shear) (Fig. 3.19). The combined effect may be attributed to shear stress due to flexure and bending stresses. The mode III (out-of-plane shear) results in cases where torsion loading is present.

Mode II Failure (In-Plane Shear)

Two ISO standard testing procedures for the determination of mode II failure in bamboo are currently in use, namely: 'bowtie test' (ISO 2004b and ISO 2004c) and the interlaminar-shear strength test. The bowtie test is described next.

Bowtie Test

This method of shear testing as illustrated in Fig. 3.20 was introduced by Janssen (1981a, b) for the determination of in-plane shear. Two bowtie plates loaded in compression on a full culm are used to induce failure at four longitudinal shear planes. The shear stress parallel to the longitudinal fibres is determined from

$$\tau_z = \frac{F_z}{4lt}$$

where

τ_z = *shear strength parallel to fibre*
F_z = *shear force*
l = *specimen length*, $l = d$
d = *outer diameter of culm*
t = *specimen thickness.*

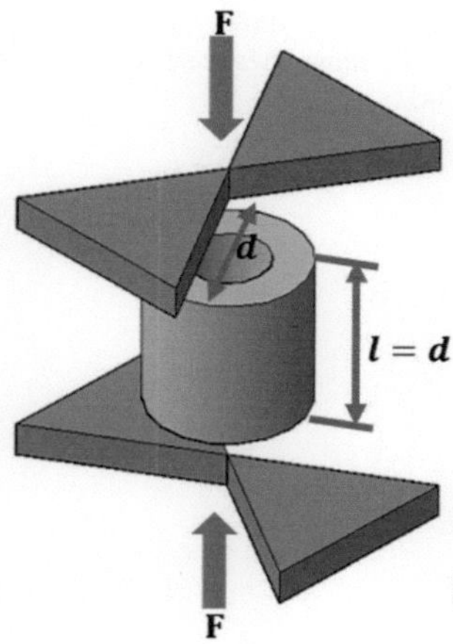

Fig. 3.20 Current test methods for in-plane shear (Janssen 1981a, b; ISO 2004)

Janssen (1981a, b) was able to establish relationships and some correlations between shear strength and fibre content, density, moisture content, position along culm, node or internode, and height. The conclusion was that there was a relationship between shear strength and all the parameters indicated above. Some parameters such as moisture content, position along culm, node or internode, and height significantly affected the shear strength.

Interlaminar-shear strength test

This test was adapted from ASTM standard D2733-70 (1976) by Moreira (1991) on the study mode-II failure. Specimens as shown in Fig. 3.21 are machined from the culm wall. Two notches at two different locations are cut to half the thickness of the specimen to create a shear plane with area A. To ensure shearing, the specimen is subjected to tensile loading that induces pure shear at its middle. It is critical that tabs are used to prevent crack or split initiation from the gripping stress caused by the gripping force. The interlaminar-shear strength is represented mathematically as:

$$\tau_{int} = \frac{F_{int}}{A_{int}}$$

τ_{int} = *interlaminar shear stress*
F_{int} = *interlaminar shear force*
A_{int} = *shear plane area*

Fig. 3.21 S-type interlaminar-shear test (Cruz 2002)

Shear Perpendicular to Fibres

Cruz (2002) designed a test for determining shear strength perpendicular to fibres in the longitudinal direction. This test is also referred to as a lap test (Fig. 3.22).

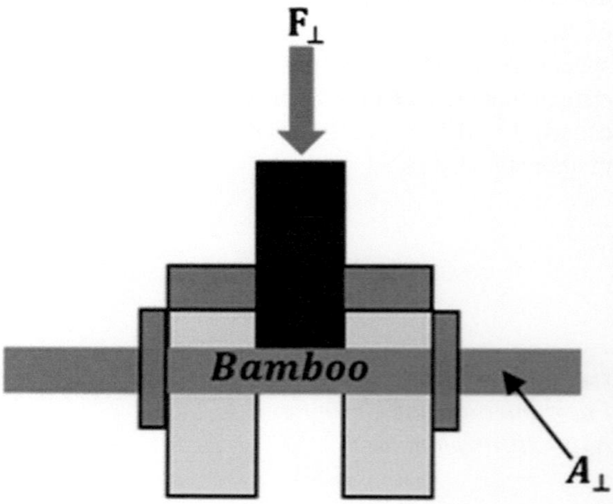

Fig. 3.22 Lap test (Cruz 2002)

The test is conducted using a simple two-plane shear configuration. The transverse shear is mathematically represented as follows:

$$\tau_{\perp} = \frac{F_{\perp}}{A_{\perp}}$$

$\tau_{\perp}$ = *Transverse shear*
$F_{\perp}$ = *shear force*
$A_{\perp}$ = *specimen area.*

Testing for the shear of bamboo is slightly complicated and presents specific challenges. A unique feature that must be present in all shear specimens is the requirement of parallel end faces. Achieving this is, however, quite challenging and time-consuming since dry bamboo naturally has high hardness. To ensure accurate results, a high degree of positional accuracy is required by utilising self-centering grips. The grips limit possible errors and significantly reduce setup time (Mitch 2009).

Shearing in bamboo is, therefore, a serious problem and the main weakness of this important plant. Having knowledge of the shear properties and strength will ensure use of an appropriate factor of safety in structural designs that use bamboo and extend the shelf life of such products or structures.

3.7.8 Bending (Flexural) Testing

The principle of loading of a simply supported beam is adapted for flexure testing. Table 3.11 shows some fundamental research undertaken on bending in bamboo.

Results obtained from maximum-fibre stress and strain are used to plot a stress-strain curve. By definition, flexural strength is the maximum stress induced in the outermost fibre of the test specimen's convex or tension side. The flexural modulus is determined from the slope of the stress-deflection curve, which may or may not have a linear region. The slope is determined by fitting a secant to the curve in cases with no linear regions.

An area of shear stress is created along the midline of the specimen during flexural testing, as tensile and compressive stresses are induced in the convex and concave sides, respectively. The shear stress is minimised by regulating the span-to depth-ratio to make certain that primary failure occurs from tensile or compressive stress.

$$\textit{span to depth ratio} = \frac{\textit{Length of outer span}}{\textit{height (depth) of specimen}} = \frac{S}{a} \tag{3.20}$$

Table 3.11 Bending

Author	Year of pub.	Method and bending strength (σ_b) and or modulus of elasticity (**E**)
H.F. Meyer and B. Ekelund	1922	**Method**: Both three-point (3-p) and four-point (4-p) tests were used **Result**: $\sigma_b = 90$ N/mm^2 (3-p) and 96 N/mm^2 (4-p)
Teodoro	1925	**Method**: three-point testing of dry *Bambusa vulgaris* (BV) *and Bambusa spinosa* (BS) were used for this study **Result**: $\sigma_b = 55$ N/mm^2 (BS) and 33 N/mm^2 (BV) E = 10,300 N/mm^2 (BS) and 18,400 N/mm^2 (BV)
Espinosa	1930	**Method**: Bending testing was carried out on dry *Bambusa spinosa* **Results**: $\sigma_b = 146$ N/mm^2 E = 14,800 N/mm^2
Trojani	1930	**Method**: Introduced a guide for determining the bending strength of bamboo: $F = \frac{100d}{3}$ Where $F =$ *applied load* (kg) $d =$ *outside diameter* (*up to three-m span only*) Also, published a nomogram for selecting the maximum load when the free span (1.5 to 4 m), culm thickness and outer diameter are known **Result**: The ultimate bending stress, (σ_b) = 20 N/mm^2, was the basis for the nomogram
Suzuki	1948	**Method**: Studied the influence of bending on the position along the culm
Glenn	1950	**Method**: Bending test is carried out to determine bending stress and modulus of elasticity on dry *Phyllostachys* **Results**: $\sigma_b = 143$ N/mm^2 E = 14,300 N/mm^2
Limaye	1952	**Method**: *Dendrocalamus strictus* was the species of bamboo used for this study A free span of 700 mm comprising a minimum of two internodes was used as the specimen configuration. This configuration implies that nodes and internodes significantly affect the test results

(continued)

Table 3.11 (continued)

Author	Year of pub.	Method and bending strength (σ_b) and or modulus of elasticity (**E**)
		Both dry and green bamboo with moisture contents 12 and 80%, respectively, were tested. Reports that the ultimate bending stress for dry bamboo was 1.5 times that for green bamboo The influence of bending along the culm was also studied **Results/Findings**: Results showed that $\sigma_b = 68$ N/mm^2 (green) and 107 N/mm^2 (dry) E = 12,000 N/mm^2 (green) and 15,600 N/mm^2 (dry) It was also observed that ultimate bending stress decreased (73–66 N/mm^2) with height, while the modulus of elasticity increased with height (10,700–13,800 N/mm^2) Bending stress was observed to increase with age
Sekhar and Bhartari	1961	**Method**: Dry *Dendrocalamus strictus* with moisture content of 12% was used for this study The effect of age on bending stress was carried out on dry bamboo **Result**: **For bamboo six years of age**: $\sigma_b = 97$ N/mm^2 E = 13,700 N/mm^2 There was no significant difference in the bending stress with age
Sekhar et al.	1962	**Method**: *Bambusa nutans* was used as the specimen for the bending test The influence of age on the bending stress for both dry and green bamboo **Result**: **For bamboo five years of age**: $\sigma_b = 100$ N/mm^2 (dry) and 79 N/mm^2 (green) E = 15,400 N/mm^2 (dry) and 13,000 N/mm^2 (green) Results for the effect of age increase were inconclusive, as some fluctuations in bending stress were observed for odd and even years
Atrops	1969	**Method**: Dry bamboo was used as the specimen for conducting the bending test **Result**: $\sigma_b = 113$ N/mm^2 Creep was found to be 65% (this result was difficult to explain)
Janssen	1981a, b	**Method**: Three-year-old full culms of *Bambusa blumeana*, with length five m, and dried to a moisture content of 12% were used for this study Janssen carried out bending tests on bamboo, subject to short- and long-term loading to determine the ultimate bending stress and modulus of elasticity. The free span for short (ST) and long (LT) terms were 3.6 and 4.5 m respectively **Result/findings: Short-term tests**:

(continued)

Table 3.11 (continued)

Author	Year of pub.	Method and bending strength (**σ_b**) and or modulus of elasticity (**E**)
		σ_b = 84 N/mm^2 E = 20,500 N/mm^2 **Long-term tests**: σ_b = 73 N/mm^2 Janssen concluded that specimens subjected to long loading periods undergo deformation 1.25 times the immediate deformation Creep was found to be 25% which could be attributed to the high percentage of cellulose

Source Janssen (1981a, b) and Arce-Villalobos (1993)

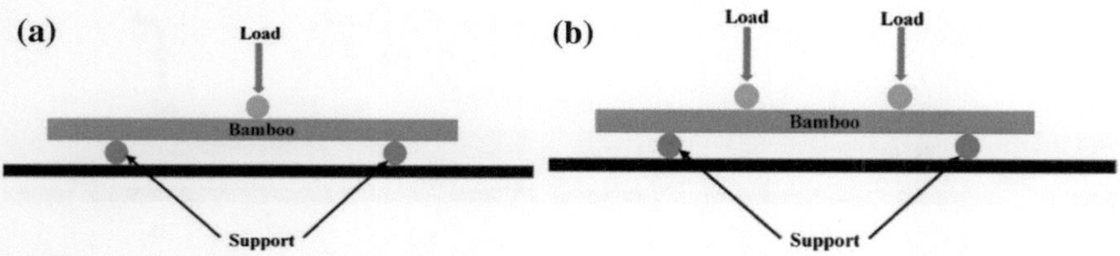

Fig. 3.23 **a** Three-point flexure test and **b** four-point flexure test

To ensure low shear stress, many engineering materials use a span-to-depth ratio of 16, while others lie within the range of 32–64.

The two types of flexure tests are the three-point and four-point flex tests which are most often used for flexible engineering materials including composites, wood, and polymers. A three-point test is characterised by a central concentrated load and a small area under uniform stress, as shown in Fig. 3.23a. By contrast, as illustrated in Fig. 3.23b, a four-point test subjects the specimen to span loading points, which results in the uniform stress area lying within the span points. A three-point bending test is mainly used for polymer specimens, while the four-point test is best suited for composite and wood materials like bamboo.

Bending testing of bamboo, in general, is difficult because of its orthotropic nature and the thickness of culm wall. Derivations and calculations are therefore difficult to obtain. From Fig. 3.24, it is observed from the bending moment and shear diagrams that failure will certainly occur at the centre for a three-point test. The failure occurs because the maximum bending moment is located at the centre, while the transverse load changes at a fast rate. The above reasons make the three-point test less reliable, with an inherently large deviation in results. The four-point bending test, however, ensures that the centre point experiences an only constant pure bending moment without the effects of transverse loading. In comparison, a four-point test is much more reliable.

3.7.8.1 Mass Per Volume

Janssen (1981a, b) reports that the ratio σ_b/G can be used to represent the relationship between the bending stress (σ_b) and the mass per volume (G). It must be noted that failure induced by bending in bamboo is mainly due to the maximum stress in the neutral layer, and hence a more practical and useful option is the shear stress (τ) to mass per volume ratio τ/G.

For a three-point test:

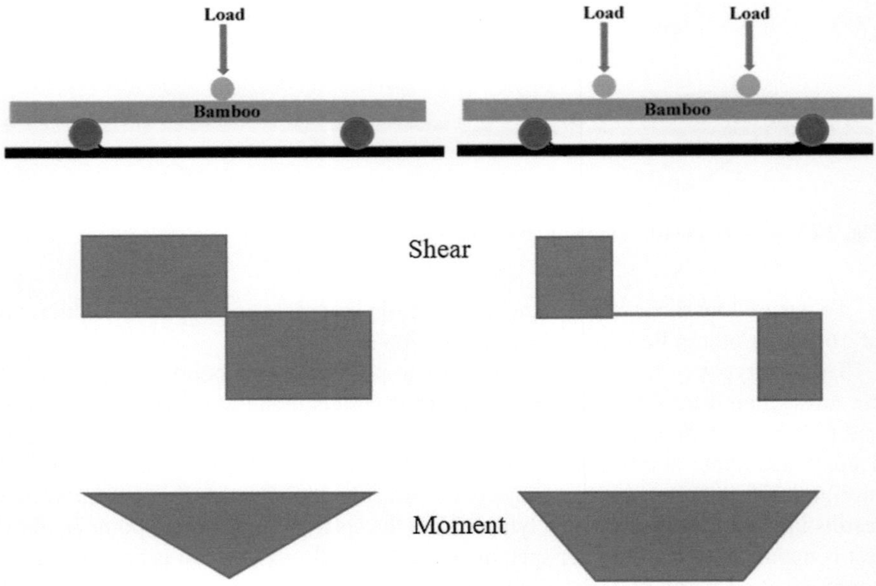

Fig. 3.24 Three-point test versus four-point test

$$\frac{\sigma_b}{\tau} = \frac{l}{d_1}$$

But

$$\sigma_b = \frac{M \times \frac{1}{2} d_2}{I}$$

$$\tau = 2 \frac{\frac{1}{2} F}{A}$$

$$M = \frac{1}{4} Fl$$

$$I = \pi \frac{\left(d_2^4 - d_1^4\right)}{64}$$

$$A = \pi \frac{\left(d_2^4 - d_1^4\right)}{4}$$

where

l = length of fee span
d_1 = inner diameter of culm
d_2 = outer diameter of culm
A = effective crossectional area of culm
F = Applied load
I = moment of inertia.

For the four-point test:

$$\frac{\sigma_b}{\tau} = \frac{2l}{3d_1}$$

$$M = \frac{1}{6}Fl$$

Tests conducted by Limaye (1952), Sekhar and Bhartari (1960, 1961), and Sekhar and Gulati (1973) confirm the conclusion made by Janssen (1981a, b) that the ratio of ultimate bending stress to the mass per volume is approximately 0.14 and 0.11 for dry and green bamboo, respectively.

3.7.8.2 Stress Distribution in Bending Test τ

While, at the neutral axis of the culm, the bending stress is zero, the shear stress is maximum, and this can result in failure. The top and bottom axis experience the maximum bending stresses due to compression and tension, respectively. Both the tensile and compressive stresses induce lateral strains that act transversely to the bamboo fibres and present another potential source of failure.

3.7.9 Torsion Testing

It is a valuable testing tool used in determining the torsional properties of materials, such as bamboo, which include ductility, shear strength, the shear modulus of elasticity, and shear-modulus rapture. Although torsional properties are analogous to those obtained by tensile testing, they are not the same. A typical example is the similarities between the stress-strain diagram and the torque-angle diagram. Products are made to undergo torsion testing in industry to check for quality, verification of design, service condition simulation, and verification of manufacturing technique. Researchers and manufacturers often use torsion testing as a critical measure of service conditions and the quality of part manufacture.

3.7.10 *Structural Equations for Bamboo*

Engineering designing requires the determination of optimal dimensions for load-carrying members. Loads that may act on structural members of wood or bamboo may be axial, bending, or a combination of axial and bending.

3.7.10.1 Structural Deformation Equations

These applied loads may result in deformations, and therefore equations for such deformations are required to aid in finding the minimum dimensions that satisfy the limits for deformation dictated by the structural design. All the equations (see Table 3.12) for deformation of wood structural parts are given as functions of dimension, load applied, modulus of elasticity, and modulus of rigidity.

3.7.10.2 Structural-Stress Equations for Bamboo

As loads are applied to wooden structural members, stresses are induced that can lead to failure of the entire structure. These stresses must be accounted for in the design process. All stresses presented in Table 3.13 are assumed to obey Hook's law and St. Venant's principle, which requires that all accepted loads and stresses must be statically equivalent to the distribution of the true stress.

3.7.10.3 Stability Equations for Bamboo

The integrity of structures built with wood can be directly linked to the stability is has. Table 3.14 presents a list of the most-used stability equations that are required to ensure the safety of such structures.

3.7.11 *Classical Laminate Theory*

Bamboo is a natural composite, and therefore the classical laminate theory may be employed in its analysis. For traditional materials, such as steel, the number of independent engineering constants required are only three (E, G or v, and α), because they are isotropic, while 12 (E_1, E_2, E_3, G_{12}, G_{13}, G_{23}, v_{12}, v_{13}, v_{23}, α_1, α_2, and α_3) constants are required to describe the elastic behaviour under mechanical and thermal loading (Carter 2001).

Table 3.12 Structural-deformation equations

Name	Equation and variables	Notes
Equations for deformation		
Axial load	$\delta = \frac{PL}{AE}$ A = cross-sectional area E = modulus of elasticity δ = change in length L = length P = axial force parallel to grain	Axial loads produce a change in length
Bending deflection of straight beams	$\delta = \frac{k_b WL^3}{EI} + \frac{k_s WL}{GA'}$ $I = \frac{bh^3}{12}$ *for rectangular cross section* $I = \frac{\pi d^4}{64}$ *for circular cross section* $A' = \frac{5}{6} bh$ *for rectangular cross section* $A' = \frac{9}{40} \pi d^2$ *for circular cross section* δ = deflection W = Acting load L = beam span k_b = constant dependence on beam loading k_s = constant dependence on support conditions I = moment of inertia A = altered area of beam E = modulus of elasticity G = shear modulus b = width of beam h = depth of beam d = diameter of beam	Values of k_b and k_s can be found in table…
Deflection of tapered beams	$\Delta_s = \frac{3WL}{20Gbh_0}$ *for uniformly distributed load* $\Delta_s = \frac{3PL}{10Gbh_0}$ *for midspan-concentrated load* h_0 = smallest beam depth Δ_s = shear deflection	
Ponding by water	$\Delta = \frac{\Delta_0}{1-S/S_{cr}}$ Δ_0 = design load deflection S = beam spacing S_{cr} = critical beam spacing	
Combined bending and axial load concentric load	$\Delta = \frac{\Delta_0}{1 \pm P/P_{cr}}$ Δ = midspan deflection under combined loading Δ_0 = beam mid-span deflection without axial load P = axial load and P_{cr} = a constant equal to the buckling load (+) = when axial load is tensile (−) = when axial load is compressive	If P is under tension, then P/P_{cr} may be neglected If P is in compression, then $P < P_{cr}$ to prevent collapse
Eccentric load	$\delta_b + \varepsilon_0 = \frac{\varepsilon_0}{1 \pm P/P_{cr}}$ δ_b = bending deflection induced at the midspan ε_0 = P's eccentricity from the centroid	Axial loading is applied to a pin-end member eccentrically to initiate deflections due to bending

(continued)

Table 3.12 (continued)

Name	Equation and variables	Notes
Equations for deformation		
Torsion	$\theta = \frac{TL}{GK}$ $K = \frac{\pi D^4}{32}$, *for circular cross sections* Or the polar moment of inertia $K = \frac{hb^3}{\emptyset}$ *for rectangular cross section* D = diameter $\emptyset$ = angle of twist [radians] T = torque L = length G = shear modulus K = cross-section shape factor h = dimension of greater cross-section b = dimension of lesser cross-section	$\emptyset$ can be derived from Fig. A

Source Soltis (1999)

Assumptions

(1) The material to be examined is made up of one or more plies (layers), each ply consisting of fibres that are all uniformly parallel and continuous across the material. The plies do not have to be of the same thickness or the same material.
(2) The material to be examined is in a state of plane stress, i.e. the stresses and strains through-the-thickness direction are ignored.
(3) The thickness is smaller than the length and width dimensions.

The convention presented is from Nettles (1994).

Convention

1. The notation used denotes the directions parallel, perpendicular, and through-the-thickness to the aligned fibres of a ply of material as 1, 2, and 3 respectively (Fig. 3.25).
2. The coordinates of the laminates are denoted by x, y (in-plane) and through the thickness z.

It must be noted that for non-isotopic materials, a minimum of two elastic constants is required to describe its stress-strain behaviour.

$$\sigma = E\varepsilon \tag{3.21}$$

where σ is stress, E is the modulus of the material, and ε is the strain.

Table 3.13 Stress equations

Stress equations		
Load	Equation and variables	Notes
Axial load Tensile stress	$f_t = \frac{P}{A}$ A = cross-sectional area f_t = tensile stress P = axial load	
Bending Straight beam stresses	$f_b = \frac{M}{Z}$ $Z = \frac{\pi D^3}{32}$, *for circular cross sections* $Z = \frac{hb^2}{6}$ *for rectangular cross section* $f_S = k\frac{V}{A}$ $k = \frac{4}{3}$ *for circular cross sections* $k = \frac{3}{2}$ *for rectangular cross section* A = cross-sectional area f_b = bending stress f_S = shear stress M = bending moment V = vertical shear force acting on cross section Z = beam section modulus	
Tapered beam stresses	$\frac{f_x^2}{F_x^2} + \frac{f_{xy}^2}{F_{xy}^2} + \frac{f_y^2}{F_y^2} = 1$ $f_x = \frac{3M}{2bh_0^2}$ $f_{xy} = f_x\, tan\theta$ $f_y = f_x\, tan^2\theta$ f_x = bending stress f_y = perpendicular stress acting on neutral axis f_{xy} = shear stress F_x, F_y, F_z = are stress derived from the design values	Norris (1950) applied the Henky-von Mises theory to wood to generate this equation
The modulus of rupture (maximum bending stress)	$\frac{R_1}{R_2} = \left[\frac{h_2L_2\left(1+\frac{ma_2}{L_2}\right)}{h_1L_1\left(1+\frac{ma_1}{L_1}\right)}\right]^{\frac{1}{m}}$ R = modulus of rupture of beam h = depth of beam L = beam span a = distance between loads m = constant dependent on the wood type	Modulus of rupture is a function of beam size and method of loading
Crack initiation criterion for beams with a slit	$\sqrt{h}\left[A\left(\frac{6M}{bh^2}\right) + B\left(\frac{3V}{2bh}\right)\right] = 1$ a = slit depth h = beam depth b = beam width M = bending moment V = vertical shear force A and B = are coefficients obtained from Fig. A	Murphy (1979) derived the fracture mechanics for analysing crack initiation in beams with a slit

(continued)

Table 3.13 (continued)

Stress equations		
Load	Equation and variables	Notes
Combined bending and axial load Concentric load	$f_b = \frac{f_{b0}}{1 \pm P/P_{cr}}$ f_b = bending stress (bending + axial load) f_{b0} = bending stress (without axial load) P = axial loading P_{cr} = buckling load (compressive axial load only) (+) = when axial load is tensile (−) = when an axial load is compressive	
Torsion	$f_s = \frac{16T}{\pi D^3}$ *for circular cross-section* $f_s = \frac{T}{\beta hb^2}$ *for rectangular cross section* f_s = shear stress due to torsion T = torque D = diameter h = greater cross sectional dimension b = lesser cross sectional dimension β = obtained from Fig. C	

Source Soltis (1999)

Therefore, using the convention above and considering the orthotropic system (Fig. 3.26):

$$\sigma_1 = E_1\varepsilon_1 \text{ OR } \sigma_2 = E_2\varepsilon_2 \tag{3.22}$$

$$Poisson's\ ratio = \nu_{12} = \frac{\varepsilon_2}{\varepsilon_1} \tag{3.23}$$

and

$$\nu_{21} = \frac{\varepsilon_1}{\varepsilon_2} \tag{3.24}$$

The relationship between Poisson's ratio and elastic modulus in each of the axes directions is:

$$\nu_{21}E_1 = \nu_{12}E_2 \tag{3.25}$$

Shear stress (τ_{12}):

$$\tau_{12} = \gamma_{12}G_{12} \tag{3.26}$$

Table 3.14 Stability equations

Stability equations		
Type	Equation and variables	Notes
Axial compression		
Long columns: Euler's formula	$f_{cr} = \frac{\pi^2 E_L}{\left(\frac{L}{r}\right)^2}$ $r = \frac{d}{4}$ *for circular cross-section* $r = \frac{b}{12}$ *for rectangular cross-section* f_{cr} = critical stress E_L = elastic modulus parallel to member's axis L = free length r = least radius of gyration d = diameter b = least cross-sectional dimension	A long column is one that buckles before the compressive stress surpasses the proportional limit
Short columns	$f_{cr} = F_c\left[1 - \frac{4}{27\pi^2}\left(\frac{L}{r}\sqrt{\frac{F_c}{E_L}}\right)^4\right]$ F_c = compressive strength	A short column is one that buckles when the compressive stress exceeds the proportional limit
Short columns: Ylinen's buckling equation	$f_{cr} = \frac{F_c + f_e}{2c} - \sqrt{\left(\frac{F_c + f_e}{2c}\right)^2 - \frac{F_c f_e}{c}}$ $f_e = \frac{\pi^2 E_L}{\left(\frac{L}{r}\right)^2}$ f_e = buckling stress c = constant lying between 0 and 1	Ylinen (1956)
Built-up columns capacity	$f_{cr} = K_f\left[\frac{F_c + f_e}{2c} - \sqrt{\left(\frac{F_c + f_e}{2c}\right)^2 - \frac{F_c f_e}{c}}\right]$ $K_f = 0.75$ *for bolts* $K_f = 0.60$ *for nails* K_f = built up stability factor and accounts for the connection efficiency	This equation is used if the applied axial load is located near the centre of the cross section and the built-up columns are properly connected
Bending		
Water ponding: Critical beam spacing	$S_{cr} = \frac{m\pi^4 EI}{\rho L^4}$ $m = 1$ *or simply supported beams* $m = \frac{16}{3}$ *for fixed end conditions* E = elastic modulus I = moment of inertia ρ = density of water (1000 kg/m^3, 0.0361 lb/in^3) L = length of beam	When beams used in the construction of roofs are spaced too wide apart with respect to their stiffness, failure under the weight of water can occur as a result of progressive deflection during steady rains
Lateral–Torsional buckling: Long beams	$f_{bcr} = \frac{\pi^4 E_L}{\alpha^2}$ $\alpha = \sqrt{2\pi}\sqrt{4\frac{EI_y}{GK}}\frac{\sqrt{L_e h}}{w}$ $EI_y = \frac{E_L d w^3}{12}$ f_{bcr} = Critical maximum bending stress α = slenderness factor E_L = elastic modulus parallel to member's axis EI_y = lateral flexural rigidity	The equations are suitable for beams of rectangular cross-section only

(continued)

Table 3.14 (continued)

Stability equations		
Type	Equation and variables	Notes
	d = depth of beam w = width of beam L_e = effective length (Table 2)	
	$\theta = \frac{TL}{GK}$ $K = \frac{hb^3}{\emptyset}$ *for rectangular cross-section* D = diameter $\emptyset$ = angle of twist [radians] T = torque L = length GK = torsional rigidity G = shear modulus K = cross-section shape factor h = dimension of greater cross-section b = dimension of lesser cross-section	
Interaction of buckling modes	$\left(\frac{f_c}{F_c}\right)^2 + \frac{f_{b1} + 6\left(\frac{e_1}{d_1}\right)f_c(1.234 - 0.234\theta_{c1})}{\theta_{c1}F'_{b1}} + \frac{f_{b2} + 6\left(\frac{e_2}{d_2}\right)f_c(1.234 - 0.234\theta_{c2})}{\theta_{c2}F'_{b2}} \leq 1.0$ $\theta_{c1} = 1 - \left(\frac{f_c}{F''_{c1}} + \frac{S}{S_{cr}}\right)$ $\theta_{c2} = 1 - \left(\frac{f_c}{F''_{c2}} + \frac{f_{b1} + 6\left(\frac{e_1}{d_1}\right)f_c}{F''_{b1}}\right)$ $F''_{c1} = \frac{0.822E}{\left(\frac{l_{e1}}{d_1}\right)^2}$ $F''_{c2} = \frac{0.822E}{\left(\frac{l_{e2}}{d_2}\right)^2}$ $F''_{b1} = \frac{1.44E}{l_e}\frac{d_2}{d_1}$ f_c = actual compressive stress f_{b1} = actual edgewise bending stress f_{b2} = actual flatwise bending stress F_c = compressive buckling strength F_{b1} = edgewise buckling strength F_{b2} = flatwise buckling strength e_1 *and* e_2 = axial compression eccentricities for edgewise and flatwise bending respectively d_1 *and* d_2 = depth of beam for edgewise and flatwise bending respectively θ_{c1} *and* θ_{c2} = moment magnification factors for edgewise and flatwise bending respectively l_e = effective length S and S_{cr} = ponding beam spacing	A combination of buckling loads, with values less than their critical values, can still cause failure

Source Soltis (1999)

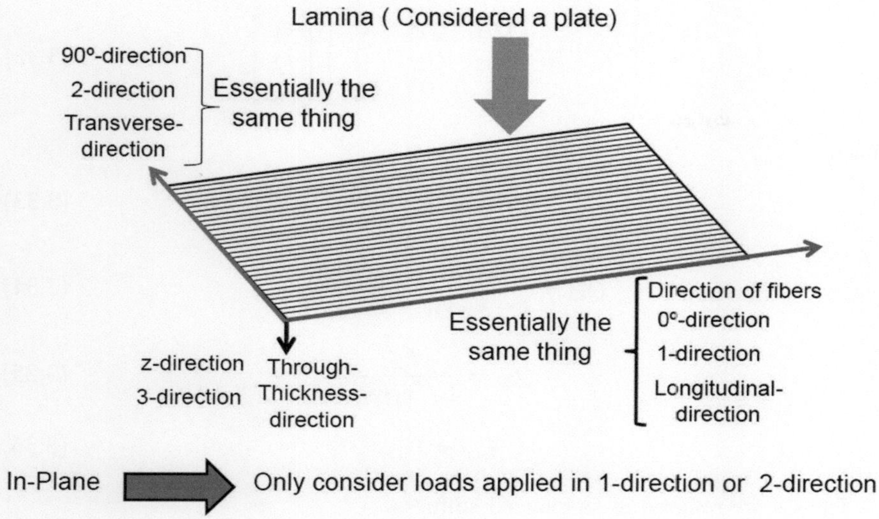

Fig. 3.25 A schematic representation of the convention used (picture from presentation by Alan T. Nettles in Mechanical Testing of Composites)

Therefore

$$\begin{bmatrix} \varepsilon_1 \\ \varepsilon_2 \\ \gamma_{12} \end{bmatrix} = \begin{bmatrix} S_{11} & S_{12} & 0 \\ S_{12} & S_{22} & 0 \\ 0 & 0 & S_{66} \end{bmatrix} \begin{bmatrix} \sigma_1 \\ \sigma_2 \\ \tau_{12} \end{bmatrix} \tag{3.27}$$

where

$$S_{11} = \frac{1}{E_1} \tag{3.28}$$

$$S_{12} = -\frac{v_{12}}{E_1} = -\frac{v_{21}}{E_2} \tag{3.29}$$

$$S_{22} = \frac{1}{E_2} \tag{3.30}$$

$$S_{66} = \frac{1}{G_{12}} \tag{3.31}$$

The 3 × 3 matrix is the compliance matrix. The inverse of the compliance matrix helps in obtaining stress as a function of strain, and the resulting matrix is called the stiffness matrix.

$$\begin{bmatrix} \sigma_1 \\ \sigma_2 \\ \tau_{12} \end{bmatrix} = \begin{bmatrix} Q_{11} & Q_{12} & 0 \\ Q_{12} & Q_{22} & 0 \\ 0 & 0 & Q_{66} \end{bmatrix} \begin{bmatrix} \varepsilon_1 \\ \varepsilon_2 \\ \gamma_{12} \end{bmatrix} \tag{3.32}$$

where

$$Q_{11} = \frac{E_1}{1 - \nu_{12}\nu_{21}} \tag{3.33}$$

$$Q_{12} = \frac{\nu_{12}E_2}{1 - \nu_{12}\nu_{21}} = \frac{\nu_{21}E_1}{1 - \nu_{21}\nu_{21}} \tag{3.34}$$

$$Q_{22} = \frac{E_2}{1 - \nu_{12}\nu_{21}} \tag{3.35}$$

$$Q_{66} = G_{12} \tag{3.36}$$

The Qs are referred to as reduced stiffness.

Summation of all forces in the 2-direction in Fig. 3.26:

$$\sum F_1 = 0 = \sigma_1 dA - \sigma_X(dA\,cos\theta)cos\theta - \sigma_y(dA\,sin\theta)sin\theta - \tau_{xy}(dA\,cos\theta)sin\theta + \tau_{xy}(dA\,sin\theta)cos\theta \tag{3.37}$$

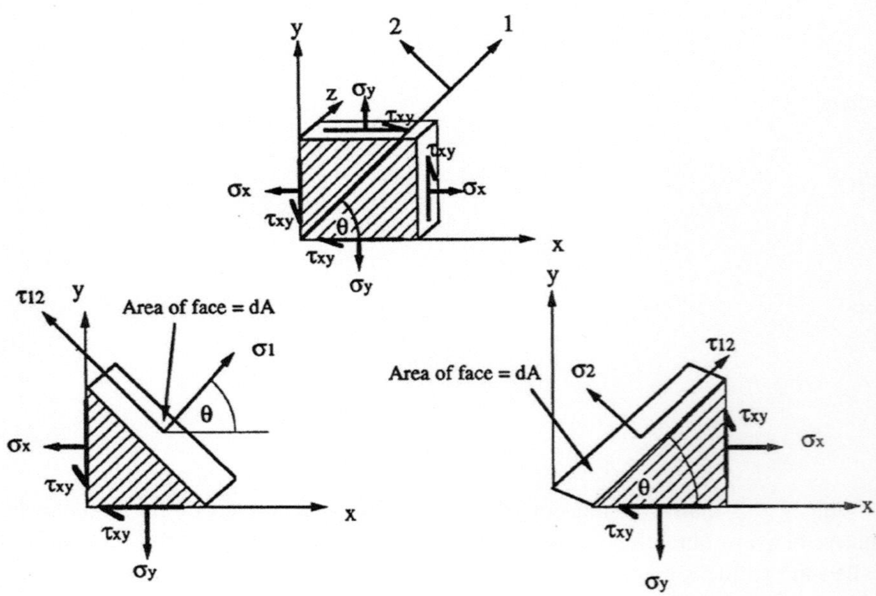

Fig. 3.26 General orthotropic lamina

$$\sum F_2 = 0 = \sigma_2 dA - \sigma_X(dA\,sin\theta)sin\theta - \sigma_y(dA\,cos\theta)cos\theta + \tau_{xy}(dA\,cos\theta)sin\theta + \tau_{xy}(dAsin\theta)cos\theta \tag{3.38}$$

$$\sum F_2 = 0 = \tau_{12} dA + \sigma_X(dA\,sin\theta)cos\theta - \sigma_y(dA\,cos\theta)sin\theta - \tau_{xy}(dA\,cos\theta)sin\theta + \tau_{xy}(dA\,sin\theta)cos\theta \tag{3.39}$$

From the simplification of Eqs. (3.37), (3.38) and (3.39):

$$\sigma_1 = \sigma_x cos^2\theta + \sigma_y sin^2\theta + 2\tau_{xy} sin\theta\,cos\theta \tag{3.40}$$

$$\sigma_2 = \sigma_x sin^2\theta + \sigma_y cos^2\theta - 2\tau_{xy} sin\theta\,cos\theta \tag{3.41}$$

$$\tau_{12} = -\sigma_x sin\theta\,cos\theta + \sigma_y sin\theta\,cos\theta + \tau_{xy}\left(cos^2\theta - sin^2\theta\right) \tag{3.42}$$

Equation (3.40), (3.42), and (3.42) can be written in the matrix form:

$$\begin{bmatrix} \sigma_1 \\ \sigma_2 \\ \tau_{12} \end{bmatrix} = \begin{bmatrix} cos^2\theta & sin^2\theta & 2sin\theta cos\theta \\ sin^2\theta & cos^2\theta & -2sin\theta cos\theta \\ -sin\theta cos\theta & sin\theta cos\theta & (cos^2\theta - sin^2\theta) \end{bmatrix} \begin{bmatrix} \sigma_x \\ \sigma_y \\ \tau_{xy} \end{bmatrix} \tag{3.43}$$

The 3 × 3 matrix above is called the transformation matrix and denoted by [T].

If it is required to transform from 1-2 to the x-y coordinate system, the inverse of [T] must be derived:

$$[T]^{-1} = \begin{bmatrix} cos^2\theta & sin^2\theta & -2sin\theta\,cos\theta \\ sin^2\theta & cos^2\theta & 2sin\theta\,cos\theta \\ sin\theta\,cos\theta & -sin\theta\,cos\theta & (cos^2\theta - sin^2\theta) \end{bmatrix} \tag{3.44}$$

Hence

For stress

$$\begin{bmatrix} \sigma_1 \\ \sigma_2 \\ \tau_{12} \end{bmatrix} = [T] \begin{bmatrix} \sigma_x \\ \sigma_y \\ \tau_{xy} \end{bmatrix} \text{ and } \begin{bmatrix} \sigma_x \\ \sigma_y \\ \tau_{xy} \end{bmatrix} = [T]^{-1} \begin{bmatrix} \sigma_1 \\ \sigma_2 \\ \tau_{12} \end{bmatrix} \tag{3.45}$$

For strain

$$\begin{bmatrix} \varepsilon_1 \\ \varepsilon_2 \\ \varepsilon_{12} \end{bmatrix} = [T] \begin{bmatrix} \varepsilon_x \\ \varepsilon_y \\ \varepsilon_{xy} \end{bmatrix} \text{ and } \begin{bmatrix} \varepsilon_x \\ \varepsilon_y \\ \varepsilon_{xy} \end{bmatrix} = [T]^{-1} \begin{bmatrix} \varepsilon_1 \\ \varepsilon_2 \\ \varepsilon_{12} \end{bmatrix} \tag{3.46}$$

The classical laminate theory is, therefore, a tool used to deduce the elastic constants required to define the elastic behaviour of materials.

3.7.12 The Effects of Water, Brine, and Acid on Bamboo

Bamboo is the preferred material for the construction of buildings and bridges in many rural communities in Africa and Asia. There are communities that build on the coast and some on water bodies. Some bridges may be constructed in either fresh water or salt water. Because of rapid industrialisation in many developing countries, a large proportion of previously pristine water bodies are becoming contaminated, due to weak environmental laws and/or weak enforcement of the laws. Some of these contaminants may be acidic in nature. This study is therefore aimed at verifying the effects on the mechanical strength and microstructure of bamboo when immersed in a meduim of water, brine, and acid.

3.7.12.1 Material and Method

B. vulgaris was used for experimentation. Samples of configurations (Figs. 3.27 and 3.28) were immersed in water, 50 g of salt diluted in 100 ml of water, and 50%

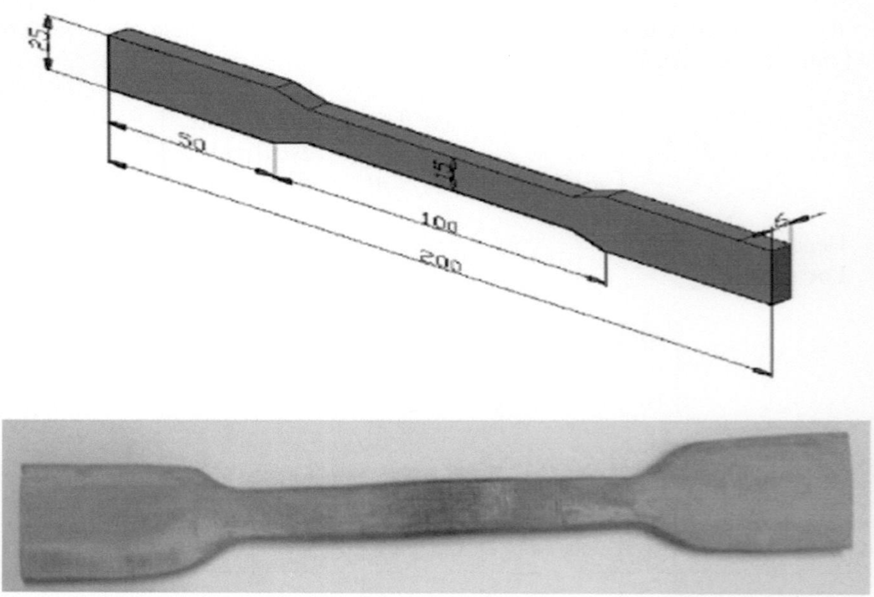

Fig. 3.27 Tensile specimen configuration

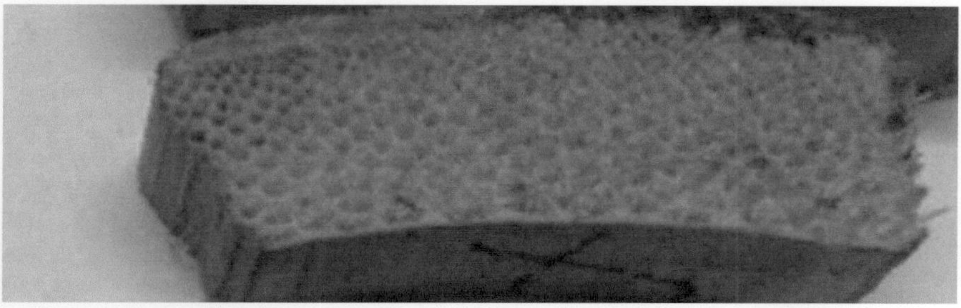

Fig. 3.28 Transverse section cut for microstructure analysis

concentrated sulphuric acid (H_2SO_4) for 24 h and air dried for 24 h. Subsequent tensile testing was conducted to determine their ultimate tensile strength. Samples for microstructure imaging were polished and viewed, using an optical microscope to generate images of 2 mm and 200 μm magnification. Dry specimens of moisture content below 12% were used as controls for experimentation.

3.7.12.2 Results and Discussion

Tensile Testing

Tensile testing revealed that the salt-treated specimen possessed the highest strength, while the worst performing was the specimen treated with acid (Fig. 3.29). From the results shown in Table 3.15 and Fig. 3.30, the mechanical properties of structures built in salty water bodies may actually improve, as the salt also serves as a preservative.

Fig. 3.29 Acid-treated sample

Table 3.15 Treatment results

Treatment	Tensile strength (Mpa)
Dry bamboo	267
Water	168
Brine	235
Acid	7

Fig. 3.30 Treatment and resulting tensile strength

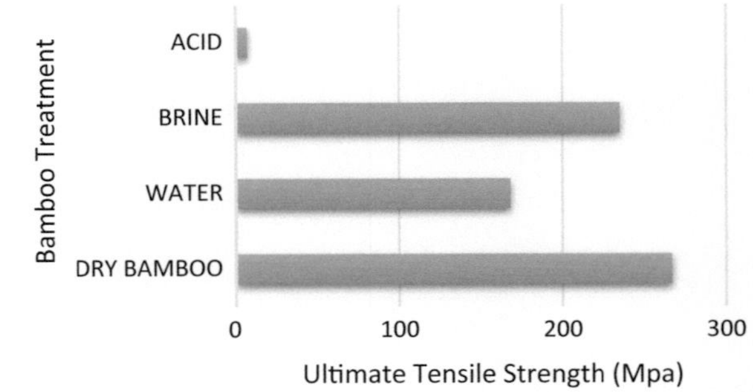

Bamboo is a lignocellulose biomass and will react to chemicals such as acids and alkaline. Fengel and Wengener (1983) described a biomass material as a composite consisting of cellulose reinforcement and a matrix of hemicellulose and lignin. The extremely low value of tensile strength for acid treatment was due to the acid causing the hemicellulose to undergo hydrolysis and become solubilized. The lignin solubilised but immediately precipitated and condensed in the acidic medium. The hydrolysis produced a black-charred look and caused the bamboo to lose its bonding integrity. This was directly responsible for the extremely low tensile strength. The implication of this result is that structures built in or around acid contaminated water will eventually lose their structural integrity.

Treated Bamboo Specimen

The samples were soaked in water, brine, and acid. The acid-treated sample underwent hydrolysis and exhibited a black charred look while the salt-treated and water-treated looked physically like the untreated sample (Fig. 3.31a–c).

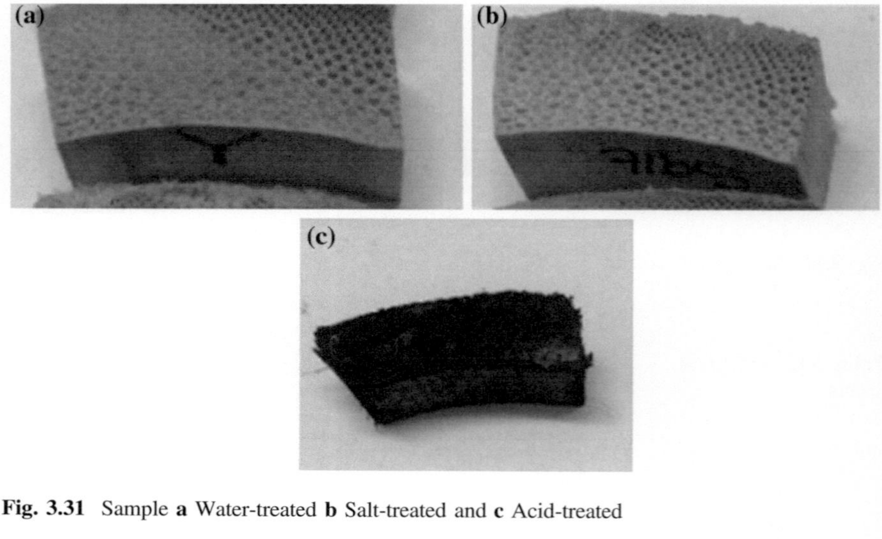

Fig. 3.31 Sample **a** Water-treated **b** Salt-treated and **c** Acid-treated

Microstructure analysis

The microstructure for all samples are shown in Fig. 3.32a–c at a magnification of 2 mm and Fig. 3.32d–i at 200 μm magnification. Figure 3.32g is the microstructure of the dry-bamboo sample, which reveals the flower-shaped vascular bundle with perfectly-formed fibre bundles, xylem, and phloem. The metaxylem supports water transfer while the metaphloem supports food transfer. The microstructure of samples soaked in water and brine were, however, slightly different (Fig. 3.32h, i). It was observed that the fibre bundles appeared fluffy, a sign of significant wetness through capillary action. The ability of bamboo to soak up liquids quickly helps facilitate processes such as liquid-perseverative treatments.

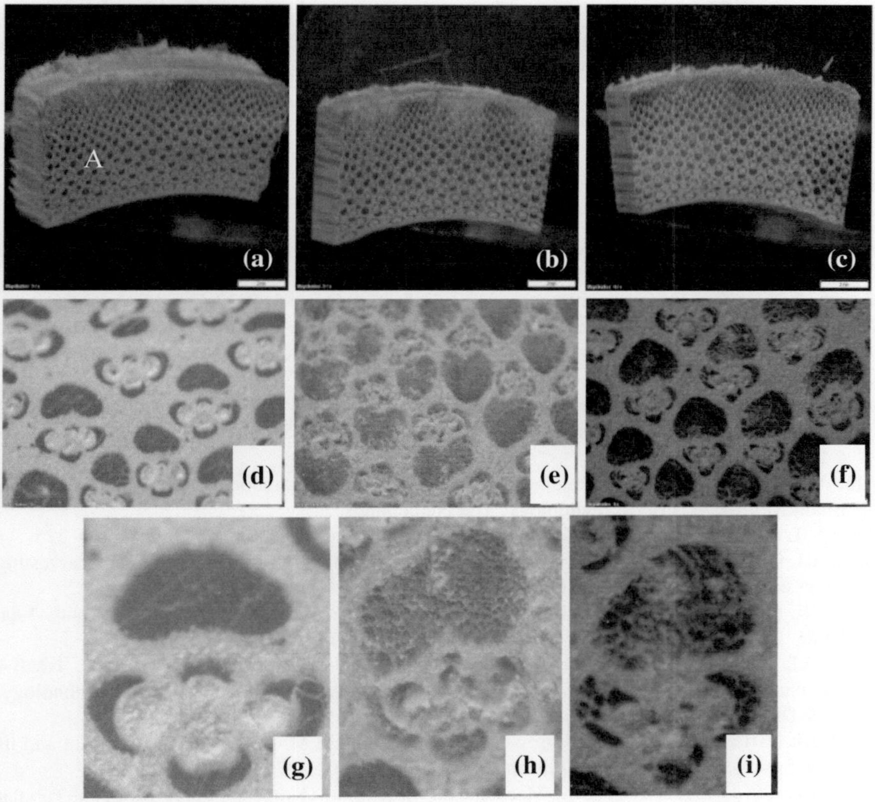

Fig. 3.32 Microstructure of **a** dry bamboo sample at magnification 2 mm, **b** water-treated sample at magnification 2 mm **c** brine-treated sample at magnification 2 mm, **d** dry-bamboo sample at magnification 200 μm, **e** water-treated sample at magnification 200 μm, **f** brine-treated sample at magnification 200 μm, **g** single vascular bundle for dry ample at magnification 200 μm, **h** single vascular bundle of water-treated sample at magnification 200 μm, **i** single vascular bundle of brine-treated sample at magnification 200 μm

The presence of moisture also led to a slight increase in the dimensions of the specimen. The drawbacks to this high-absorption capability of bamboo are that it attains high saturation levels quickly that are detrimental to cell life. For this reason, bamboo cannot survive in swampy or waterlogged grounds.

3.7.12.3 Summary

This study has determined the effects on bamboo when subjected to liquid environments such as fresh water, salt water, and acid solutions. The results revealed that bamboo structures retained their tensile-strength integrity in both fresh and salt water. However, all structural integrity was lost when subjected to an acidic environment. The microstructure of all samples tested showed that bamboo stems exhibited a strong affinity for moisture absorption, which is excellent for preservative treatment but detrimental for survival in waterlogged lands.

References

Amada S, Lchikawa Y, Munekata T, Nagase Y, Shimizu H (1997) Fiber texture and mechanical graded structure of bamboo. Compos B 288:13–20

Amada S, Untao S (2001) Fracture properties of bamboo. Compos B 32:451–459

Amada S, Munekata T, Nagase Y, Ichikawa Y, Kirigai A, Yang Z (1996) The mechanical structures of bamboos in viewpoint of functionally gradient and composite materials. J Compos Mater 30(7):800–819

American Society of Testing Materials (ASTM) (1997) Standard guide for moisture conditioning of wood and wood-based materials. Annual Book of ASTM Standards Des. D 493391. Vol – 4.10. Philadelphia

Arce-Villalobos OA (1993) Fundamentals of the design of bamboo structure. Ph.D. Thesis, Eindhoven University of Technology, The Netherlands

Atrops JL (1969) Elastizitaetund festigkeit von bambusrohren. Der Bauingeniur 44:220–225

Banik RL (1997a) The edibility of shoots of Bangladesh bamboos and their continuous harvesting effect on productivity. Bang J For Sci 26(1):1–10

Banik RL (2000) Silviculture and field-guide to priority bamboos of Bangladesh and South Asia. BFRI, Chittagong, pp 1–187

Banik RL (2010) Biology and silviculture of muli (*Melocanna baccifera*) bamboo. NMBA (National Mission on Bamboo Applications), TIFAC, Department of Science and Technology, New Delhi, pp 1–237

Banik RL (2015) Morphology and growth. In: Liese W, Kohl M (eds) Bamboo: the plant and its uses. Springer International Publishing, Switzerland, pp 43–87

Carter RH (2001) Characterizing the Mechanical Properties of Composite Materials Using Tubular Samples. Phd Thesis, Virginia Polytechnic Institute and State University

Chatterjee RN, Raizada MB (1963) Culm-sheaths as aid to identification of bamboos. Indian For 89(11):744–756

Chen YD, Qin WL (1985) The chemical composition of ten bamboo species. In: Rao AN (ed) Recent research on bamboo. Proceedings of the International Bamboo Workshop, Hangzhou, China, 6–14 October. Chinese Academy of Forestry, Beijing, China; International Development Research Center, Ottawa, Canada, pp 110–113

Cox FB, Geymayer HG (1969) Expedient reinforcement for concrete for use in S. E. Asia. Monograph

Cruz MLS (2002) Caracterizaão fisica e mecânica de colmos inteiros do bambu da espécie phyllostachys aurea: comportamento à flambagem. M.S. thesis, Pontifícia Universidade Católica, Rio de Janeiro, Brazil (in Portuguese)

De Vos V (2010) Bamboo material properties and market perspectives. Thesis report of Larenstein University, pp 19–48

Duff CH (1941) Bamboo and its structural use. In: The Engineering Society of China session 1940–41, Paper ICE. Institute of Civil Engineers of Shanghai, pp 1–27

Espiloy ZB (1987) Physico-mechanical properties and anatomical relationships of Philippine bamboo. In: Rao AN, Dhanarajan G, Sastry CB (eds) Recent research on bamboos. Proceedings of the 3rd international bamboo workshop, Hangzhou, China, 6–14 October 1985. Chinese Academy of Forestry, Beijing, China; International Development Research Centre, Ottawa, Canada, pp 257–264

Espinosa JC (1930) Bending and compressive strengths of the common Philippine bamboo. Philippine J Sci 41(2):121–135

Fengel, D and Wegener G (1983) Wood: Chemistry, Ultrastructure, Reactions. Walter de Gruyter, Berlin

Forest Products Laboratory (1999) Wood handbook: wood as an engineering material. FRS catalogue no. 7269, U.S. Department of Agriculture

Ghavami K (1995) Ultimate load behavior of bamboo-reinforced lightweight concrete beams. Cement Concr Compos 7:281–288

Glenn HE (1950) Bamboo reinforcement in portland concrete. Engineering Experiment Station, South Carolina, 4

Gnanaharan R, Janssen JJA, Arce O (1994) Bending strength of guadua bamboo—comparison of different testing procedure. INBAR working paper no. 3. Kerala Forest Research Institute (KFRI), India and International Development Research Centre (IDRC), Canada

Gross M (2009) Green shoots. Curr Biol 19(9):345–6. http://www.ncbi.nlm.nih.gov/pubmed/23638598. Accessed 10 May 2016

Grosser D, Liese W (1971) On the anatomy of Asian bamboos, with special reference to their vascular bundles. Wood Sci Technol 5:290–312

Grosser D, Liese W (1973) Present status and problems of bamboo classification. J Arnold Arboretum 54(2):293–308

Hamdan H (2004) Characterization of mechanical properties and behaviour of *Giganthochloa scortechinii* for structural application. Ph.D. Thesis. University of Wales, Bangor

Hamdan H, Hill CAS, Zaidon A, Anwar UMK, Latif A-M (2007) Equilibrium moisture content and volumetric changes of *Gigantochloa scortechinii*. J Trop For Sci 19(1):18–24

Hamdan H, Zaidon A, Tamizi MM (2009) Mechanical properties and failure behaviour of *Gigantochloa scortechinii*. J Trop For Sci 21(4):336–344

Harries KA, Sharma B, Richard M (2012) Structural use of full culm bamboo: the path to standardization. Int J Archit Eng Constr 1(2):66–75

INBAR (1999) An international model building code for bamboo. The International Network on Bamboo and Rattan, Beijing

Islam MN, Hannan MO, Lahiry AK (2002) Effect of age and height positions of borak, jawa and mitinga bamboo on their physical properties. J Timber Dev Assoc India 48:16–22

ISO (2004) ISO 22157-1: bamboo—determination of physical and mechanical properties—part I: requirements. International Standards Organization, Geneva

Jamaludin K, Jalil HA, Ashari AJ, Latif AM (1992) Variation in specific gravity of 1-, 2- and 3 years old *Gigantochloa scortechinii* (Buluh Semantan). In: Wan Razali WM, Aminuddin M (eds) Proceeding, First National Bamboo Seminar, Forest Research Institute Malaysia, Kuala Lumpur, Malaysia, pp 182–185

Janssen JA (1981a) The relationship between mechanical properties and the biological and chemical composition of bamboo. In: Higuchi T (ed) Bamboo production and utilization

Kyoto, Japan: Proceedings of XVIII IUFRO World Congress, Kyoto, 1981, Kyoto University, pp 27–32

Janssen JJA (1981b) Bamboo in building structures. Ph.D. Thesis, Eindhoven University of Technology, The Netherlands

Jiang X, Li Q (1982) Preliminary study on vascular bundles of bamboo native to Sichuan. J Bamboo Res 1:17–21

Kamthai S (2003) Alkaline sulphite is pulping and ECF-bleaching of sweet bamboo (*Dendrocalamus asper* Backer). M.S. Thesis, Kasetsart University, Thailand

Karamchandani KP (1959) Role of bamboo as a constructional material. In: Proceedings of a Symposium on Timber and Allied Products, Building Organisation of India, New Delhi, 18–22 May 1959, pp 430–435

Kamruzzaman M, Saha SK, Bose AK, Islam MN (2008) Effect of age and height on physical and mechanical properties of bamboo. J Trop For Sci 20(3):211–217

Kelemwork S (2008) Testing of two Ethiopian Bamboo species for value add products. Retrieved from http://www.eabp.org.et/Publications/present/TestingofTwoEthiopianBamboo.pdf

Kumar S, Shukla KS, Dev I, Dobriyal PB (1994) Bamboo preservation techniques—a review. INBAR Technical Report No. 3. International Network for Bamboo and Rattan, New Delhi; Indian Council of Forestry Research and Education, Dehra Dun, p 59

Latif A-M, Liese W (2002a) Culm characteristics of two bamboos in relation to age, height and site. In: Kumar A, Ramanuja Rao IV, Sastry C (eds) Bamboo for sustainable development. Proceedings of the 5th international bamboo congress and 6th international bamboo workshop, San José, Costa Rica, 2–6 November 1998. VSP and INBAR, pp 223–233

Latif A-M, Liese W (2002b) The moisture content of two Malaysian bamboos in relation to age, culm height, site and harvesting month. In: Kumar A, Ramanuja Rao IV, Sastry C (eds) Bamboo for sustainable development. Proceedings of the 5th international bamboo congress and 6th international bamboo workshop, San José, Costa Rica, 2–6 November 1998. VSP and INBAR, pp 257–268

Li X (2004) Physical, chemical, and mechanical properties of bamboo and its utilization potential for fiberboard manufacturing. M.S. Thesis, Chinese Academy of Forestry

Li XB, Shupe TF, Peter GF, Hse CY, Eberhardt TL (2007) Chemical changes with the maturation of the bamboo species *Phyllostachys pubescens*. J Trop For Sci 19(1):6–12

Liese W (1980a) Anatomy and properties of bamboo. http://www.studymode.com/essays/Bamboo-1302838.html. Accessed 6 Oct 2015

Liese W (1980b) Anatomy of bamboo. In: Bamboo research in Asia. Proceedings of a workshop at International Development Research Centre, Singapore, Ottawa, pp 161–164

Liese W (1985a) Bamboos, biology, silvics, properties, utilization. Deutsche Gesellschaft fiir Technische Zu sanimenarbeit (GTZ), Eschborn, p 126

Liese W (1985b) Anatomy and properties of bamboo. In: International Bamboo Workshop (INBAR), 6–14 October, China

Liese W (1998) The anatomy of bamboo culms. International Network for Bamboo and Rattan (INBAR), Technical report no. 18, Beijing, China

Liese W, Grosser D (1972) Untersuchungen zur Variabilita¨t der Faserla¨nge bei Bambus. Holzforschung 26(6):202–211

Liese W, Ding Y (2003) Structure and functions of the nodes in bamboo. In: Shanmughavel P, Peddappaiah RS, Liese W (eds) Recent advances in bamboo research. Scientific Publishers (India), pp 193–199

Liese W, Grover PN (1961) Untersuchungen über den Wassergehalt von indischen Bambushalmen. Ber Deut Bot Ges 74:105–117

Liese W, Latif AM (2000) The starch content of two Malaysian bamboos in relation to age, culm height, site and harvesting month. In: Proceedings of the 21st IUFRO world congress, Kuala Lumpur, Malaysia, 7–12 August 2000. Poster Abstracts 3:261

Liese W, Tang TKH (2015) Morphology and growth. In: Liese W, Kohl M (eds) Bamboo: the plant and its uses. Springer International Publishing, Switzerland, pp 227–253

Limaye LD (1952) Strength of bamboo (*Dendrocalamus strictus*). Indian For 78:558–575

Lo TY, Cui HZ, Leung HC (2004) The effect of fiber density on strength capacity of bamboo. Mater Lett 58:2595–2598
Long C, Liu Y, Xue J, Li D (2003) A bamboo germplasm collection for community development in Central Yunnan, China. J Bamboo Rattan 2(1):1–2
McClure FA (1966) The bamboos: a fresh perspective. Harvard University Press, Cambridge, pp 1–347
McLaughlin EC (1979) A note on the strength of Jamaica grown bamboo. Wood Fibre 11(2): 86–91
Meyer HP, Ekelund B (1922) Tests on the mechanical properties of bamboo. In: 87 The Engineering Soc of China Session 1922–1923, vol 22 paper 7, Present in library E.T.H. Switzerland, pp 141–169
Meyer HF, Ekelund B (1923) Tests on the mechanical properties of bamboo. The Engineering Society of China, session 1922–1923 22(7):141–169
Mitch DR (2009) Splitting Capacity Characterization of Bamboo Culms. Undergraduate Thesis, University of Pittsburgh
Mitch D, Harries KA, Sharma B (2010) Characterization of splitting behavior of bamboo culms. ASCE J Mater Civ Eng 22:1195–1199
Moreira LE (1991) Desenvolvimento de estruturas treliadas espaciais de bamboo. Ph.D. Thesis, Pontifícia Universidade Católica, Rio de Janeiro, Brazil (in Portuguese)
Muller I, Rebelo C (2015) Bamboo Worldwide: The current market and future potential. http://www.ecoplanetbamboo.com/files/bamboo_worldwide.pdf. Accessed 24 Oct 2016
Murphy JF (1979) Using fracture mechanics to predict failure of notched wood beams. In: Banff AB (ed) Proceedings of first international conference on wood fracture, BC: Forintek Canada Corporation, Vancouver, 14–16 Aug. 1978, 159: 161–173
Nettles AT (1994). Basic Mechanics of composite Plates. Nasa Reference Publication
Norris CB (1950) Strength of orthotropic materials subjected to combined stresses. WI: U.S. Department of Agriculture, Forest Service, Forest Products Laboratory, Rep. 1816. Madison
Nor Aziha MA, Azmy HM (1991) Preliminary study on the four Malaysian commercial bamboo species. India Bull 1(2):6–10
Ota M (1950) Studies on the properties of bamboo stem (part 10). On the relation between the tensile strength parallel to the grain and the moisture content of splint bamboo. Bull Kyushu Univ For 23:155–164
Ota M (1953) Studies on the properties of bamboo stem. Part 9: On the relation between compression strength parallel to grain and moisture content of bamboo splint. Bulletin Kyushu University Forestry, 22: 87–108
Prawirohatmodjo S (1988) Comparative strengths or green and air dry bamboo. In: Proceedings of the International Bamboo Workshop, Cochin, India, 14–18 November 1988, pp 218–222
Puls J (1992) α-Glucuronidases in the hydrolysis of wood xylans. In: Visser J, Beldman G, Kusters-van Someren MA, Voragen AGJ (eds) Xylans and xylanases. Progress in biotechnology, vol 7. Elsevier, Amsterdam, pp 213–224
Qisheng Z, Shenxue J, Yongyu T (2001) Industrial utilization on bamboo. Technical report no. 26. International Network for Bamboo and Rattan (INBAR), Beijing
Rangqui C, Kuihong W (1987) Study on physico-mechanical properties of *Phyllostachys fibriligula*. J Bamboo Resour, China 6(3):20–25
Ratanophat C (2004) Sustainable management and utilisation from bamboo. Final technical report. Project: PD 56/59 Rev. 1(I) Promotion of the utilisation of bamboo from sustainable sources in Thailand. Royal Forest Department and International Tropical Timber Organization, Bangkok, Thailand
Razak AO, Latif AM, Liese W, Naron N (1995) Planting and utilisation of bamboo in Peninsular Malaysia. Forest Research Institute Malaysia (FRIM), Kuala Lumpur, Malaysia. Technical Pamphlet no. 118
Razak W, Janshah M, Hashim WS, Shirley B (2007) Morphological and anatomical characteristics of managed natural bamboo stands *Gigantochloa scortechinii*. J Bamboo Rattan 6(2)

Resource Centre for Natural Fibre Crafts Project (2015) Resource Centre for Natural Fibre Crafts Project funded by Office of Development Commissioner (Handicrafts) Ministry of Textiles, Govt. of India. http://www.fibreresource.org/raw-materials/bamboo. Accessed 24 Sept 2016

Richard MJ (2013) Assessing the Performance of Bamboo Structural Components. Doctoral Dissertation, University of Pittsburgh

Sattar MA, Kabir MF, Battacharjee DK (1994) Effect of age and height position of muli (Melocanna baccifera) and borak (Bambusa balcooa) bamboos on their physical and mechanical properties. In: Proceedings of the 4th international bamboo workshop bamboo in Asia and the Pacific, Chiangmai, Thailand, 27–30 Nov 1991. International Development Research Centre, Ottawa, Canada, and Forestry Research Support Program for Asia and the Pacific, Bangkok, Thailand, pp 183–187

Schroder S (2011) Bamboo identification: bamboo stem anatomy. http://www.guaduabamboo.com/identification/bamboo-stem-anatomy. Accessed 7 Feb 2016

Sekhar AC, Bhartari RK (1960) Studies on strength of bamboo: a note on its mechanical behavior. Indian For, 296–301

Sekhar AC, Bhartari RK (1961) A note on strength of dry bamboo (*Dendrocalamus strictus* from Madhya Pradesh. Indian For 99:611–613

Sekhar AC, Rawat BS (1957) Strength test on bamboo: *Bambusa nutans*. Indian For Leaflet (147)

Sekhar AC, Rawat BS, Bhartari RK (1962) Strength of bamboo: *Bambusa nutans*. Indian For, 67–73

Sekhar AC, Gulati AS (1973) A note on the physical and mechanical properties of Dendrocalamus strictus from different localities. Van Vigyan (J Soc Indian For), 11(3/4):17–22. Sekhar AC, Rawat BS, Bhartari RK (1962) Strength of bamboos: Bambusa Nutans. Indian Forester, 67–73

Sharma A (1990) Bamboo reinforced concrete beams. In: Proceedings of the IABSE Symposium, Brussels, pp 677–682

Sharma B (2010) Performance based design of bamboo structures. Ph.D. thesis, University of Pittsburgh, Pittsburgh, United States

Sharma B, Gatóo A, Bock M, Ramage M (2015) Engineered bamboo for structural applications. Constr Build Mater 81(66–7):3

Shaw A (2012) Characterization of engineered bamboo for buildings. Indian Institute of Technology, Delhi

Siau JF (1995) Wood: influence of moisture on physical properties. Department of Wood Science and Forest Products, Virginia Polytechnic Institute and State University

Simpson W, Tenwolde A (1999b) Physical properties and moisture relations of wood. Wood handbook-wood as an engineering material. Gen. Tech. Rep. FPL-GTR-113. Forest Products Laboratory, U.S. Department of Agriculture, Forest Service, Madison

Soeprayitno T, Tobing L, Widjaja EA (1988) Why the Sundanese of west java prefer slop-inhabiting *Gigantochloa pseudoarundinacea* to those growing in the valley. In: Proceedings of the International Bamboo Workshop, Cochin, India, 14–18 November 1988, pp 215–217

Soltis LA (1999) Structural analysis equations. In: wood handbook—wood as an engineering material. Forest Products Laboratory, U.S. Department of Agriculture, Forest Service, Forest Products, Wisconsin

Sotela J (1992) Propiedades fisicas y mecanicas de la Gaudua de Costa Rica. Seminario Centroamericano Usos del Bambu, Pocora, Limon, Costa Rica

Sulthoni A (1989) Bamboo: physical properties, testing methods and means of preservation. In: Bassali AV, Davies WG (eds) Proceedings of a workshop on design and manufacturing of bamboo and rattan furniture, Jakarta, Indonesia, 3–14 Mar 1989, pp 1–15

Suzuki Y (1948) Distribution of specific gravity and bending strength along the culm. Bull Tokyo Univ For 36:188

Tamolang FN, Lopez FR, Semana JA, Casin RF, Espiloy ZB (1980) Properties and utilization of Philippine bamboos. In: Lessard G, Chouinard A (eds) Bamboo research in Asia. IDRC, Ottawa, pp 189–200

Taylor D, Kinane B, Sweeney C, Sweetnam D, O'Reilly P, Duan K (2014) The biomechanics of bamboo: investigating the role of nodes. Wood Sci Technol 49(2):345–357. http://doi.org/10.1007/s00226-014-0694-4

Teodoro AL (1925) A preliminary study of the transverse strength of structural bamboos. Agric Eng 6(11):266

Torres LA, Ghavami K, Garcia JJ (2007) A transversely isotropic law for the determination of the circumferential young's modulus of bamboo with diametric compression tests. Lat Am Appl Res 37(4):255–260

Trojani F (1930) Passerelles en bambous. Bull. Econ. Indo-China, London, pp 369B–373B

Vena PF, Brienzo M, del García-Aparicio MP, Rypstra T (2013) Hemicelluloses extraction from giant bamboo (*Bambusa balcooa* Roxburgh) prior to kraft or soda-AQ pulping and its effect on pulp physical properties. Holzforschung 67:863–870

Wakchaure MR, Kute SY (2012) Effect of moisture content on physical and mechanical properties of bamboo. Asian J Civ Eng (Build Hous) 13(6):753–763

Wu S, Wang H (1976) Studies on the structure of bamboos grown in Taiwan. Bulletin of the National Taiwan University 16:79

Xiu-Xin L, Liu-Keqing W (1985) A study on the physico-mechanical properties of culm wood of Phylostachys glauca of Shandog. J Bamboo Res 4(2)

Ylinen A (1956) A method of determining the buckling stress and the required cross-sectional area for centrally loaded straight columns in elastic and inelastic range. Publication of the International Association for Bridge and Structural Engineering, 16, Zurich

Yusoff MNM, Kadir A, Mohamed AH (1992) Utilization of bamboo for pulp and paper and medium density fiberboard. In: Mohd WR, Mohamad AB (eds) Proceeding of the seminar towards the management, conservation, marketing and utilization of bamboos. FRIM, Kuala Lumpur, pp 196–205

Chapter 4
Bamboo as Fuel

Abstract The rising cost of, and pressure on reserves of, fossil fuels such as petroleum, coupled with sustainability and environmental issues, has led to research in alternative biofuels that are greener and more sustainable. One such biofuel is bamboo. Its fast growth rate is one of the main driving factors for its consideration as a sustainable option. In this study, a review is conducted of the production of charcoal, pellets, ethanol, and methane gas from bamboo. Bamboo has been found to be capable of generating commercial quantities of ethanol and methane gas. However, pre-treatment process optimisation is still required. Experimentation on bamboo pellets to determine their average calorific value and the fluent-as emission levels associated with their combustion is reported. The average calorific value (CV) observed for bamboo pellets is of 17,650 J/kg, which satisfies the minimum requirement for commercial use. Emissions from the combustion of pellets were within acceptable limits. Bamboo is therefore considered a viable alternative to fossil fuels.

4.1 Introduction

Fossil fuels are the main sources of energy for heating and cooking activities in the developed world. The situation is however different in the developing world, where as high as 50–90% of consumed energy is from wood fuels such as firewood, charcoal, pellets, biogas (methane), and ethanol (FAO 2010).

Bamboo has many applications; however, its use as a fuel is probably gaining the greatest attention from industry and research. High fuel prices, coupled with environmental and sustainability concerns, have fueled the drive towards alternative sources of energy such as solar, wind, hydro, and biofuels.

This search for sustainable fuel resources has led to innovative ways to generate energy from several types of biofuels of which bamboo is one. Bamboo is a special fuel source, whether it is mature or immature, dry, or has high-moisture content. In this chapter, several techniques for transforming bamboo into fuel are discussed. These techniques include the pyrolysis of bamboo for the production of charcoal

E.T. Akinlabi et al., *Bamboo*, DOI 10.1007/978-3-319-56808-9_4

and pellets and the use of pretreatment to produce ethanol and methanol. Using bamboo charcoal seems to be a viable alternative fuel to forest wood for most developing countries in Africa because it has a faster growth rate and is less expensive to use. Experiments were conducted on bamboo pellets to assess their calorific and gas-emission values.

4.2 Bamboo Charcoal

Global trend Sub-Saharan Africa is a major producer and consumer of charcoal. In 2007, 8 billion USD was the estimated value placed on charcoal production in this region. Projections show an upward trend to 12 billion USD by 2030 (Sander et al. 2011). Brazil is, however, the largest producer of charcoal. A comparison of the energy content for normal firewood and charcoal is 16 and 30 MJ/kg, respectively. In 2011, the global estimate for the production of charcoal was 47 million metric tonnes (FAO Stat 2013), with global production having increased 11% per annum since 2003. Similar increases in charcoal production have been reported in many of the developing countries, especially in Africa and Asia. Based on a UNDP and WHO (2009) report on charcoal and firewood, see Fig. 4.1 for usage in some West African countries.

Another study, conducted by the Food and Agriculture Organisation of the United Nations (FAO) Statistics (FAO 2009), states that the consumption of wood fuel increased from 20,678,000 m^3 in 2004 to 35,363,400 m^3 in 2008, while the consumption of wood charcoal likewise increased from 752,000 m^3 to 1,477,700 m^3 during the same period (Fig. 4.2) (FAO 2009).

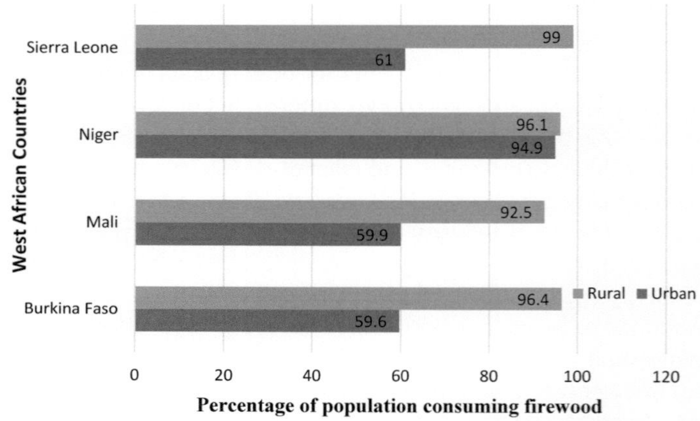

Fig. 4.1 Firewood consumption in some West African countries. *Adapted from* UNDP and WHO (2009)

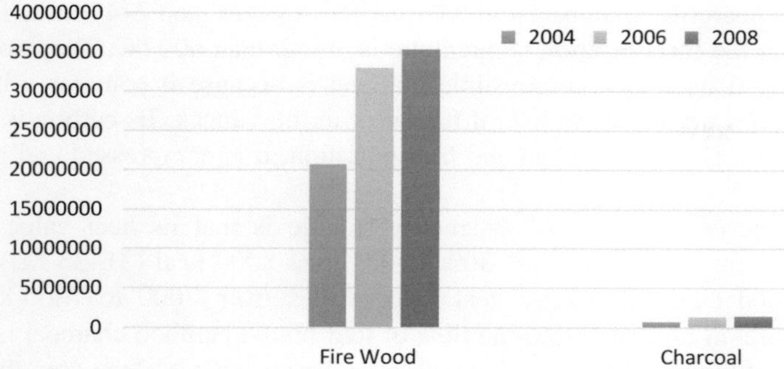

Fig. 4.2 Consumption of wood fuel and wood charcoal (*Source* FAO 2009)

Production The production of bamboo charcoal basically involves three main steps, namely: the cutting of bamboo culm into smaller sections, heating to carbonize them in kilns under controlled conditions, and, lastly, quality control and packaging (Fig. 4.3).

The kiln is the most important piece of equipment in the production of bamboo charcoal. To achieve optimal charcoal production strict, temperature control is a requirement. The bamboo undergoes about five discrete steps at varying temperatures during the carbonizing process within the kiln. These periods include:

1. Smoking at a temperature between 60 and 100 °C
2. Drying at 100–150 °C
3. Pre-carbonizing at 150–300 °C
4. Carbonizing at 300–450 °C
5. Refining at 450–1000 °C (Fu and Chen 2009).

To categorise bamboo charcoals, the technical specifications for quality assessment are sorted into two grades. The calorific values for first grade and second grade bamboo charcoal is $\geq$ 33,400 J/g and $\geq$ 32,300 J/g, respectively.

Fig. 4.3 Culm charcoal

Bamboo charcoal contains about 85% carbon, 7% gas, and 2–4% ash. The pH is above 7.0, if the carbonisation temperature is higher than 600 °C. The two primary reasons why it used as a combustible material is because it contains a high heat value and does not carry the risk of fungal or termite attack. Its carbon is in a fine granular form. For easy storage and transportation, it is compressed and packaged into cubes.

An important indication of its energy storage is that its heat value rises, at combustion temperatures of 500–800 °C, to 7400–8000 kcal (31–33 MJ), in contrast to wood charcoal, whose heat value varies from 7000 to 7800 kcal (29–33 MJ). With an estimated burning time of four hours, bamboo charcoal is suitable for use in a fireplace. Bamboo charcoal has a larger inner surface area than wood charcoal and thus has better absorptive properties. The density of bamboo charcoal at a temperature of 300 °C reaches 565 kg/m^3 and at 1000 °C climbs to 720 kg/m^3 (Fu and Chen 2009).

4.3 Bamboo Pellet Biofuel

The evolution in lifestyle and continued global population increase, together with the industrialisation of rising global powers in Asia, South America and Africa, have led to an increase in global energy demand. In 2015, the global usage of bio-power rose by 5% to 106.4 GW. The leading biomass electricity generating countries, in descending order, are USA (69 TWh), Germany (50 TWh), China (48 TWh), Brazil (40 TWh), and Japan (36 TWh), United Kingdom and India. Wood pellets, which tend to possess high energy densities, have been produced by Western Europe, Russia, and North America, and, since 2014, by Japan and the Republic of Korea, as the primary global suppliers. The development of quality standards and sustainable certification for the wood-pellet industry were instituted by the Sustainable Biomass Partnership in 2015 (REN21 2016).

Kumar and Chandrashekar (2014) investigated five bamboo species, namely: *Dendrocalamus strictus*, D. *brandisii*, D. *stocksii*, *Bambusa bambos,* and B. *Balcooa* to determine their fuel characteristics. The study revealed that bamboo's calorific value ranges from 18.7 to 19.6 MJ/kg^{-1}. Liu et al. (2012) studied the influence of moisture content and particle size on the physical and combustion properties of pellets. The research revealed that moisture content had a statistically significant effect on physical properties, such as length, diameter, and unit density. However, particle size made no significant difference.

The current global trend shows steady gravitation towards greener and more sustainable energy resources, achieved mainly through the adoption of alternative energy sources, such as solar, wind, hydro, biomass, ocean thermal, and geothermal, replacing petroleum and other fossil fuels The wide spectrum of biomass available makes it a preferred option because of its accessibility, affordability, and option as the only carbon-based alternative energy source (Faizal et al. 2010). Bamboo is a biomass material that is attracting a great deal of attention because it is a sustainable resource and very easy to cultivate and maintain. Because of bamboo's constituents, which are mainly hemicellulose, cellulose, and lignin, it is well-suited for use as a bioenergy resource. Comparing its growth rate with most trees, it recovers at a rate of 2–3 years while most forest trees take 60 years; it is biodegradable; frugal in water consumption; thrives in diverse climates and soil conditions; and, as a grass, it does not necessarily require replanting (Steinfeld 2001; Yu et al. 2003; Afrin et al. 2009). Although both raw bamboo and bamboo charcoal can be used, densification into pellets is a better alternative as it provides improved flow properties, reduction in waste, and is easier to store and transport (Adapa et al. 2006). The steps for manufacturing pellets commercially are graphically listed in Fig. (4.4).

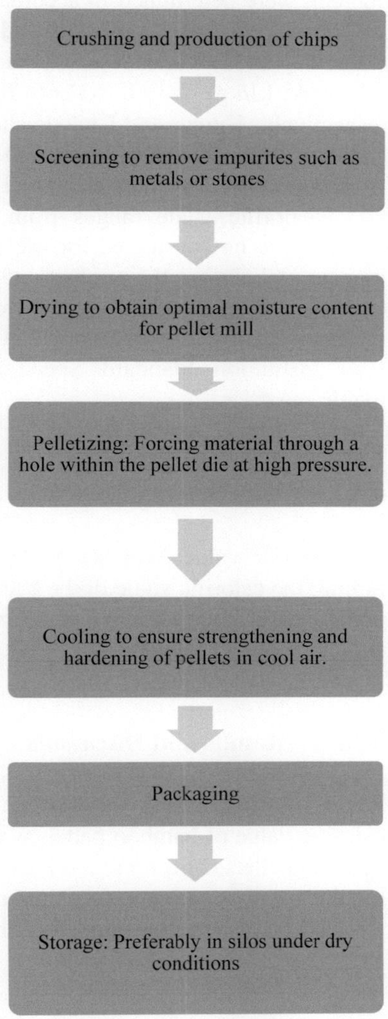

Fig. 4.4 Seven steps in pellet production. *Source* Anyang Gemco Energy Machinery (2016)

4.3.1 *Characterisation of Bamboo Pellets*

See Table (4.1).

4.3.2 *Calorific Value of Bamboo Pellets*

In this study, the calorific-value determination and the gas-analysis test were carried out on commercial pellets of *Bambusa vulgaris* manufactured in Ghana. The physical properties, including pellet dimensions, unit and bulk densities, and ash content, were determined in accordance to EN 14961-2.

The calorific values of soft and hardwoods were determined by Telmo and Lousada (2011). The research measured the calorific values of (19,660.02–20,360.45 kJ/kg) and (17,631.66–20,809.47 kJ/kg) for softwood and hardwoods, respectively. Kumar and Chandrashekar (2014) investigated five bamboo species, namely: *Dendrocalamus strictus*, D. *brandisii*, D. *stocksii*, *Bambusa bambos,* and B. *Balcooa,* to determine their fuel characteristics. The study revealed that bamboo's calorific value ranges from 18.7–19.6 MJ·kg-1. For any biomass pellet (Fig. 4.5) to be considered for use in Europe, which is the leading consumer and pioneer global user, the product must meet the minimum requirement for the gross calorific value of 17,500 J/g for commercial pellets stipulated in DIN 51731 (German standard) or EN 14961-2 (European). The pellets must also meet the Pellet Fuel Institution Standard Specification for Residential/Commercial Densified requirement.

4.3.2.1 Material and Method

The calorific-value Analyser Network station Cal2k (Fig. 4.6) was used in determining the calorific value of the bamboo pellets. Three samples of crushed bamboo pellets, each having a mass of 0.5 g, were used for this experiment. Table 4.2 is a summary of the calorific values obtained.

4.3.2.2 Results and Discussion

Table 4.2 summarises the results: after three experimental runs, the average calorific value of bamboo pellets was found to be 17,650 J/g. Table 4.3 shows that bamboo pellets conform to the DIN 51731 and EN 14961-2 standard calorific value required as the minimum for commercialisation of wood pellets. Figure 4.7 graphs the temperature profile for 100 counts during the combustion.

Table 4.1 Characterisation of bamboo pellets

Characterisation	Equation and variables	Notes
Dimension measurement	d = pellet dimension (mm) l = length (mm)	Each pellet is measured using Vernier callipers
Unit density	$\rho = \frac{M}{v}$ $v = \frac{\pi}{4d^2 l}$	Calculation: The volume of the individual pellets is obtained from the previously-stated dimensions
Unit density	D = pellet dimension (mm) l = pellet length (mm) M = mass of individual pellet (g) v = unit volume (cm^3) ρ = unit density (g/cm^3)	The pellets are then weighed and used in calculating the unit density, using Eqs. 1 and 2
Bulk density	$\rho_b = \frac{M_{total}}{v_c}$ ρ_b = bulk density (g/cm^3) M_{total} = total mass of pellet (g) v_c = standardised container's volume (cm^3)	A standardised container with known volume is filled to the top with pellets and weighed. Bulk density can then be determined by calculating the ratio of the mass obtained to the volume of the standardised container
Moisture content	$MC = \left(\frac{M_1 - M_2}{M_2}\right) \times 100\%$ MC = moisture content (%) M_1 = sample initial mass (g) M_2 = sample final mass (g)	In compliance with EN 14774-1, the determination of moisture content requires using 300 g of bamboo pellets heated in a drying oven at a standard temperature of 105 °C for eight hours. After this duration, the pellets are removed from the dryer, cooled at room temperature, and then weighed
Pellet fine	$p_f = \left(\frac{M_1 - M_2}{M_1}\right) \times 100\%$ p_f = pellet fine (%) M_1 = sample initial mass (g) M_2 = sample final mass (g)	The pellet fine is determined according to **Pellets Fuel Institute Standard Specification for Residential/Commercial Densified requirements** by using a wire-screen sieve of dimension 3.17 mm (1/8 in). Selected pellets are divided into groups, and then their initial mass is determined and poured onto the sieve. The sieve is tilted ten times from side to side. The final mass of the pellet is observed and used in calculating the pellet fine
Pellet durability	$p_d = 100 - \left(\frac{M_1 - M_2}{M_1}\right) \times 100\%$ p_d = pellet durability (%) M_1 = sample initial mass (g) M_2 = sample final mass (g)	Mass of selected samples was determined and then placed on a vibrating standard screen sieve of dimension 3.17 mm for approximately ten minutes. The final mass is recorded and used to calculate the pellet durability

Source Liu et al. (2012)

Fig. 4.5 Commercially produced bamboo pellets

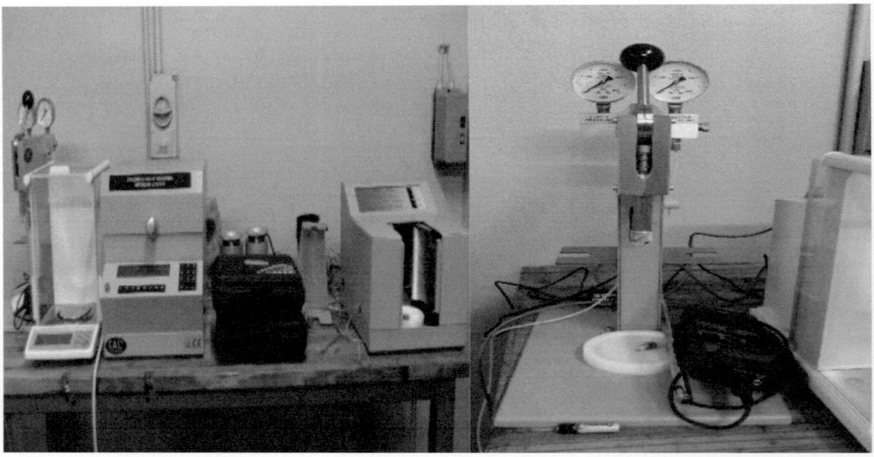

Fig. 4.6 The calorific value analyser network station Cal2 k

4.3.3 Gas-Analysis Test for Bamboo

As early as the 1970s, Europe has lead in the densification of wood for use as pellets for heating purposes Vinterback (2004). Due to several stimulus initiatives and tax benefits to manufacturers of biomass products for electricity and heating in Europe, an annual growth of 25% has been forecast for the industry (Vinterback 2004; Malisius et al. 2000). Pelletization is receiving a great deal of attention because it

Table 4.2 Calorific value of bamboo pellets

Number of tests	Mass (g)	Calorific Value (J/kg)
1	0.494	17,630
2	0.498	17,780
3	0.486	17,550
		Averaged (CV) = 17,650

Table 4.3 Comparison between DIN 51731/EN 14961-2 standard and bamboo-pellet calorific values

DIN 51731/EN 14961–2	Bamboo pellet
17,500 (J/g)	17,650 (J/g)

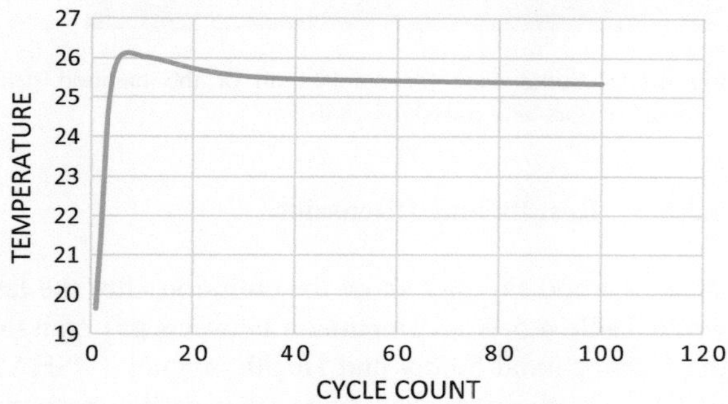

Fig. 4.7 Temperature profile for 100 counts during the combustion

provides clean burning, can be manufactured from wood waste (sawdust or wood shavings), and is very convenient to use. European countries such as Germany, Norway, and Sweden are using wood pelletization for heating as a means to reduce GHG emissions, by incentivising the industry with various programmes (Magellia et al. 2009).

Gas-analysis testing is conducted to determine the gases emitted during combustion of solid or liquid fuels, such as wood, briquettes, pellets, petroleum-based, and organic-based fuels.

4.3.3.1 Material and Method

The *Bambusa vulgaris* species of bamboo was used in the manufacture of the pellets. A mass of 7.78 g of the pellets was crushed and placed in the combustion chamber of a stove (Fig. 4.8). The Gas Analyser E5500 probe was used to take the measurement of the emitted volatile organic gases, such as CO_2, CO, SO_2, SO, NO, NO_2, and NOX.

(a) **(b)**

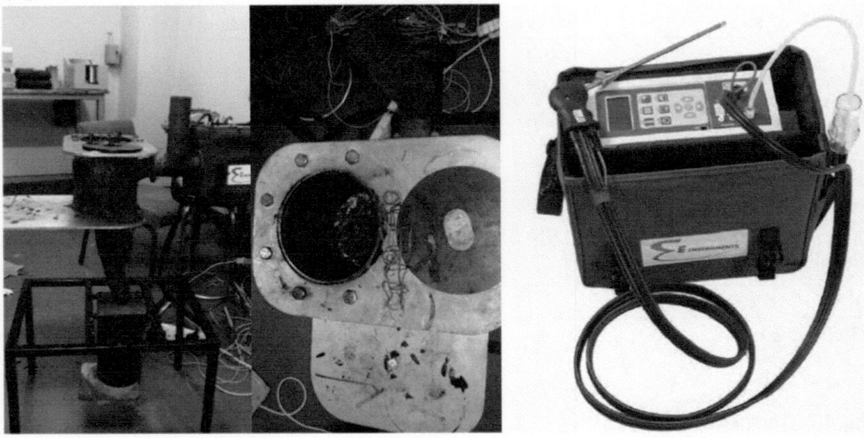

Fig. 4.8 **a** Stove used for combustion of the bamboo pellets. **b** E5500 Portable industrial combustion gas and emissions analyser

4.3.3.2 Results and Discussion

Table 4.4 and Fig. 4.9 show the emission after the last glow of the bamboo pellet, while Table 4.5 is a comparison between gas emissions from bamboo pellets and the Occupational Safety and Health Agency (OSHA) occupational exposure standards. It is observed that, aside from carbon monoxide, all other emissions were below and thus conformed to the OSHA limits. A compilation of flue-gas emissions from other fuel sources are presented in Table 4.6. The comparison of flue-gas emission concentrations for natural gas, fuel oil, coal, eco-fuel briquettes, and bamboo pellets shows that bamboo pellets as an alternative fuel is less toxic than other fossil fuels.

The emission of oxygen during combustion ranged from a maximum of 20.85% at initiation to 19.82% at completion (Fig. 4.10). The high level of oxygen indicates that complete combustion was ensured. The high concentration of carbon dioxide implies complete combustion occurred, and therefore carbon monoxide was oxidised to the less toxic carbon dioxide, as shown in Fig. 4.11. It is better to have higher levels of carbon dioxide, which is a greenhouse gas that can be easily monitored and managed than having higher levels of more toxic emissions such as NO and SO_2 (Fig. 4.12). To avoid the harmful effect of carbon dioxide and carbon

Table 4.4 Emissions from bamboo pellets after final glow

	O_2 (%)	CO (ppm)	CO_2 (%)	NO (ppm)	NO_2 (ppm)	SO_2 (ppm)
Emissions	19.82	897.42	0.24	10.87	0	1.08

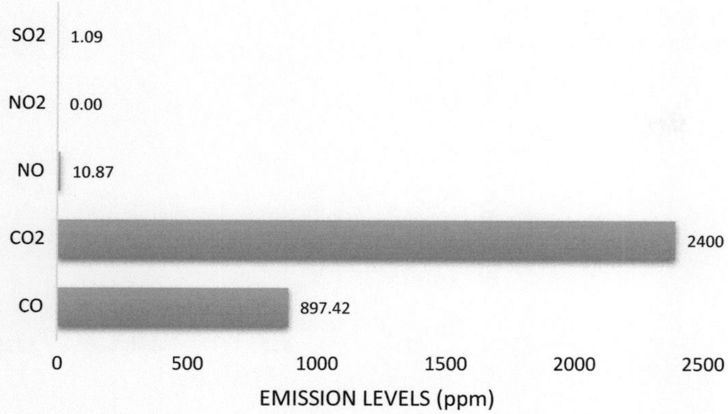

Fig. 4.9 Flue gas emission levels from bamboo pellets

Table 4.5 Comparison of toxic emissions from bamboo pellets and accepted levels from OSHA

	CO (ppm)	CO_2 (ppm)	NO (ppm)	NO_2 (ppm)	SO_2 (ppm)
Pellet	897.42	2400	10.87	0	1.08
OSHA limit	50–200	5000	25	5	5

Table 4.6 Flue-gas quality comparison of bamboo-pellet emissions with other available fuels

Emissions	OHSA limit (ppm)	Natural gas	Fuel oil	Coal	Eco-fuel briquettes	Bamboo pellets
(O_2)	–	2–3%	2–6%	7%	12.8%	19.82%
(CO)	50–200	70–110 ppm		5579 ppm	74 ppm	897 ppm
(CO_2)	5000	10–12%	12–14%	10.6%	21.3%	2400 ppm
(NO)	25			1%	1.34 ppm	10.87 ppm
(NO_2)	5			1%	2.73 ppm	0
(SO_2)	5			>2000	3.67 ppm	1.08 ppm

Note Values for OSHA, Natural gas, fuel oil, coal, and eco-fuel briquettes are referenced from Pilusa et al. (2013)

monoxide, proper ventilation is required for dilution if indoor applications are intended.

4.3.3.3 Summary

Bamboo pellets can be recommended as a viable alternative fuel source for heating and electricity generation because its gross calorific value of 17,650 J/g meets the minimum requirement of 17,500 J/g recommended by commercial pellets by DIN

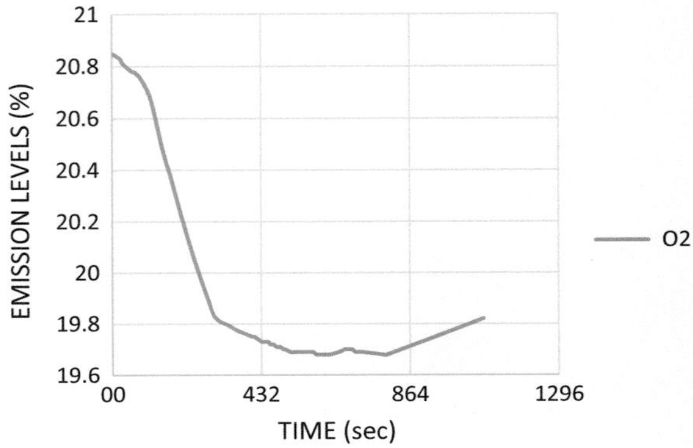

Fig. 4.10 Oxygen emissions during combustion of bamboo pellets

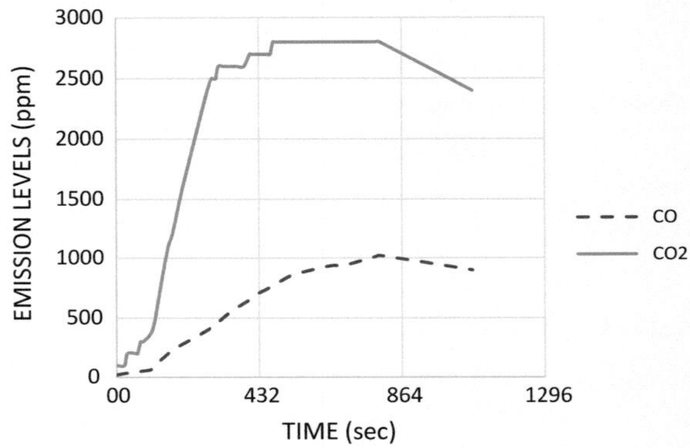

Fig. 4.11 CO and CO_2 emissions during combustion of bamboo pellets

51731 (German standard) or EN 14961-2 (European). Also, bamboo pellets during combustion emit less toxic gases when compared with fossil fuels.

4.4 Production of Ethanol from Bamboo

Bamboo has many applications in the energy industry, mainly for the production of pellets, charcoal, and briquettes for heating and electricity. The crusade to limit or reduce the emission of greenhouse gases from fossil fuels has led to an intensified move towards research into non-food plants for use in the development and production of biofuels (Sticklen 2008; Sims et al. 2010). In efforts to achieve the above goals, lignocelluloses have been identified as the only potential feedstock capable of replacing fossil fuels (Kim and Dale 2004; Gray et al. 2006), and bamboo falls into this category. Many researchers have conducted studies to produce ethanol from bamboo as a potential source of biogas (Kobayashi et al. 2004) and as an alternative renewable and sustainable fuel resource (Zhang et al. 2007; Shimokawa

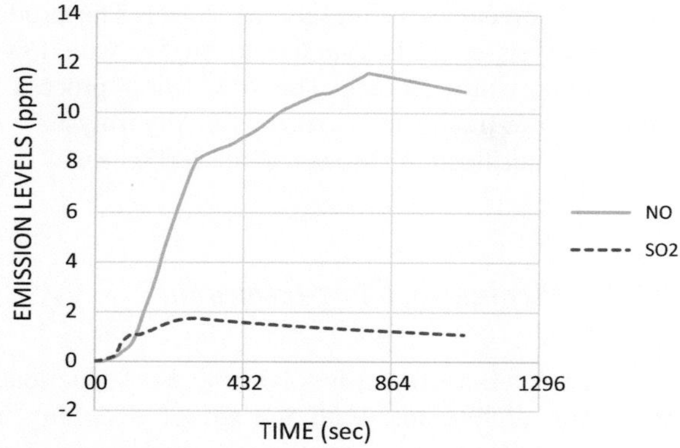

Fig. 4.12 NO and SO_2 emissions during combustion of bamboo pellets

et al. 2009; Yamashita et al. 2010; Sun and Lin 2010; Sathitsuksanoh et al. 2010). Bamboo is an attractive proposition because it possesses a growth rate higher than most forest trees and consists of very high holocellulose content of about 70% of its dry base (Sun et al. 2011). These characteristics of bamboo make it a viable biofuel source for production of bioethanol which can satisfy the energy needs of a nation, while being relatively environmentally friendly (Farrell et al. 2006).

Bamboo as a lignocellulose plant consists of very complex structures, namely: cellulose, hemicelluloses, and lignin (Rubin 2008; Xu 2010). The cellulose and the lignin are linked by the hemicelluloses to constitute a rigid cell-wall structure. This linkage is responsible for inhibiting hydrolysis.

Hemicellulose consists of monosaccharides that include hexoses (mannose, glucose, and galactose), pentoses (xylose and arabinose), hexoses, and sugar acids, which form polysaccharide complexes in nature. Hydrolysis of hemicelluloses is easily accomplished because of the low molecular weight, short lateral chains of branches, and amorphous structures (Fengel and Wegener 1983; Saha 2003; Hedriks and Zeeman 2009; Li et al. 2010).

Cellulose consists of d-glucose subunits connected by structurally amorphous and crystalline-(1,4)-glycosidic bonds that form a linear polymer. Cellulose strains come together to form fibrils that are also linked by very weak hydrogen bonds (Laureano-Perez et al. 2005; Hendriks and Zeeman 2009).

Lignin consists of three monomeric (phenylpropane units) alcohols namely, trans-p-coumaroyl alcohol, trans-p-coniferyl alcohol, and trans-p-sinapyl alcohol—all linked together to form a heterogeneous polymer which is amorphous in nature (Fengel and Wegener 1983; Kirk-Othmer 2001; Hendriks and Zeeman 2009; Li et al. 2010; Wen et al. 2011).

In comparison with other biomass sources, bamboo's proportions of cellulose and lignin are relatively high and low, respectively, making it a great potential source for production of bioethanol. There are three main determining factors essential to biomass conversion into sugars capable of fermentation, namely: the quantity of convertible polysaccharides, p-treatment methods, and enzymatic hydrolysis efficiency (Kirk-Othmer 2001). The production of bioethanol requires the hydrolysis of carbohydrates to sugars with low molecular weight and then fermentation into ethanol. The production process involves four major stages, namely: pre-treatment, enzymatic hydrolysis, fermentation and product separation/purification (Mosier et al. 2005a, b).

4.4.1 Mechanical Pre-treatment

Mechanical pre-treatment involves the size reduction of lignocellulosic biomass to particulate dimensions, with the aid of a milling machine. During the milling process, reduction in crystallinity also occurs. The advantages of mechanical pre-treatment are that it assures shearing, an increment in the specific surface area, and a reduction in the degree of polymerisation (DP). These three variables contribute to a 5–25% increase in the total hydrolysis yield and a 23–59% decrease in the time required for technical digestion (Palmowski and Muller 1999; Delgenés et al. 2002; Hartmann et al. 1999). It must, however, be noted that little to no effects on yield and rate of hydrolysis occur when milled particulates have dimensions below 40 mesh (Chang and Holtzapple 2000). An increase in the hydrolysis rate and ethanol yield, while hydrolysis inhibitors such as furfural and HMF (hydroxymethylfurfural) are absent when milling, is used for pre-treatment. The main disadvantage is the uneconomically high energy consumption associated with milling. Due to continually increasing energy costs, researchers (e.g. Cowling and Kirk 1976; Ramos 2003; and Fan et al. 1987) have concluded from extensive research that milling or mechanical pre-treatment is not economically viable.

4.4.2 Thermal Pre-treatment

Thermal pre-treatment involves the heating of lignocellulosic biomass within the range 150–180 °C. The solubilisation of hemicellulose begins at temperatures above 180 °C through an exothermal reaction (Beall and Eickner 1970; Domansky and Rendos 1962). The hydrolysis of hemicellulose and the formation of acids takes place during thermal pre-treatment. The production of the acids seems to be an added advantage since they induce and catalyse further hydrolysis (Gregg and Saddler 1996). Several extensive researches have shown that, beyond 160 °C, solubilisation occurs in both hemicellulose and lignin, which are inhibitory to hydrolysis (Liu and Wyman 2003; Zhu et al. 2004, 2005). It was observed that there was a high probability of the formation of toxic phenolic and heterocyclic compounds (e.g. vanillin, vanillin alcohol, furfural, and HMF) from soluble hemicellulose and lignin at elevated temperatures (Ramos 2003). At a temperature above 250 °C, pyrolysis occurs, and complete inhibition of ethanol takes place (Brownell et al. 1986; Laser et al. 2002).

4.4.2.1 Steam Pre-treatment/Steam Explosion (ST/SE)

High-pressure steam of about 240 °C is used for pre-treatment of biomass in a vessel for a set time, after which the steam is evacuated and the biomass subjected to fast cooling. Cellulose is made accessible during steam/steam-explosion pre-treatment because solubilisation of hemicellulose takes place. The digestion of enzymes during biomass steam pre-treatment was observed to increase six-fold (Grous et al. 1986). There is, however, the potential risk for the formation of fermentation inhibitors such as furfural, phenol, and HMF compounds during the production of ethanol (Hindriks and Zeeman 2009).

4.4.2.2 Liquid Hot Water (LHW)

Liquid hot-water pre-treatment involves the use of hot water with a pH between 4 and 7 to solubilise hemicellulose, to ensure better accessibility of cellulose for enzymatic hydrolysis. The objective of choosing pH lying between 4 and 7 is to limit the formation of inhibitors such as monosaccharides (Kohlmann et al. 1995; Mosier et al. 2005a; Weil et al. 1997).

The main advantage of LHW over steam pre-treatment is that concentrations of solubilised hemicellulose and lignin precipitants are lower (due to the high water input), and therefore the potential risk of formation of inhibitors is severely reduced. The resulting outcome, as Weil et al. (1998) observed, is a two-to-five increase in yield in biomass enzymatic hydrolysis.

4.4.3 Chemical Pre-treatment

Chemical pre-treatment can be described as using acids and alkaline oxidising agents to change the physico-chemical composition of lignocellulose biomass.

4.4.3.1 Acid Pre-treatment

Acid pre-treatment is the use of strong or dilute acids to improve anaerobic digestion at ambient temperature. During pre-treatment with acid, the hemicellulose undergoes hydrolysis, becomes solubilized, and can undergo further hydrolysis to produce furfural, HMF, monomers, and volatile compounds which are not suitable for ethanol production (Fengel and Wegener 1983; Ramos 2003). Also, lignin solubilises but immediately precipitates and condenses in the acidic medium. Sulfuric or nitric acids are often used for pre-treatment. It must be noted that the dilute acids tend to induce less-pronounced solubilised hemicellulose and precipitation of solubilised lignin. Therefore, in ethanol production, strong acids should be avoided.

4.4.3.2 Alkaline Pre-treatment

Solvation and saphonication take place during alkaline pre-treatment and are the first reactions that occur to induce swelling, which directly increases the specific surface area of the biomass, making it more prone to enzyme and bacteria attacks. A distinct feature of this pre-treatment process is the consumption of the alkali by the biomass itself, and therefore the only the residual alkali left behind is involved in the reaction (Gossett et al. 1982; Fengel and Wegener 1983). Gossett et al. (1982) concluded that lime was better for alkaline pre-treatment than sodium hydroxide. The hemicellulose and lignin solubilise during alkaline pre-treatment, and some inhibitory compounds can influence fermentation and are hence detrimental to the production of ethanol. These characteristics of alkaline pre-treatment make it a less preferred option for the production of ethanol (Hendriks and Zeeman 2009).

4.4.3.3 Thermal Pre-treatment/Alkaline Pre-treatment

Alkaline and thermal pre-treatments can be combined to produce an improved process that is much more efficient in ethanol production. Chang et al. (2001) experimented using lime ($Ca(OH)_2$) at a temperature range of 100–150 °C and found that softwoods, after pre-treatment, required washing to limit possible inhibition by solubilised lignin before proceeding to the later stages of enzyme saccharification and fermentation of ethanol production. Lime is the most preferred choice because it is cheaper than other alkalis, safer to handle, and can be reclaimed by using kiln technology on the calcium carbonate that results from pre-treatment (Gandi et al. 1997; Chang et al. 1998).

4.4.3.4 Oxidation Pre-treatment

Oxidation pre-treatment involves making the cellulose accessible by the removal of hemicellulose and lignin by the use of oxidising agents, such as peracetic acid, and hydrogen peroxide applied to the biomass in the production of ethanol. Most oxidants, however, are not selective during oxidation, and therefore some hemicellulose and cellulose may be lost, while inhibitors from the oxidation of lignin may form (Hon and Shiraishi 2001). Peracetic acid has been found to be very selective in oxidising lignin, which can lead to a cellulose increment during hydrolysis, from 6.8 without treatment to approximately 98% after treatment—when 21% of the peracetic acid is utilised in pre-treatment (Teixeira et al. 1999).

4.4.3.5 Ammonia Pre-treatment

The ammonia pre-treatment (AFEX pre-treatment) has produced very positive results for the production of ethanol. Alizadeh et al. (2005) conducted an extensive study on the use of ammonia and revealed that performing pre-treatment at ambient and 120 °C temperatures, required durations of 10–60 days and some minutes, respectively, resulting in increased yields of six times for enzyme hydrolysis and 2.5 times for ethanol yield.

4.4.4 Hydrolysis and Fermentation

Many researchers [e.g. Chang and Holtzapple (2000), Koullas et al. (1992), Laureano-Perez et al. (2005), Puri (1984), Grethlein (1985), Grous et al. (1986), Thompson et al. (1992), Gregg and Saddler (1996), Grethlein (1985) and Palonen et al. (2004)], have reported that the inhibiting factors to enzymatic hydrolysis are cellulose crystallinity, the degree of polymerization (DP), moisture content, available surface area, and lignin content. The use of enzymes for the hydrolysis of lignocellulosic biomass is the most common approach because its sugar yield is high; however, the enzymatic saccharification process requires expensive pre-treatment methods. Fu and Mazza (2011) observed that the sugar recovered from enzyme saccharification of cellulose was high, and other researchers, such as Li et al. (2010, 2011), showed that the sugar concentration was very low for industrial purposes. Iranmahboob et al. (2002) and Binder and Raines (2010) have shown that using enzymes for hydrolysis is currently not attractive for industrial applications in the production of ethanol.

A study conducted by Clausen and Gaddy (1993) revealed that sulfuric-acid hydrolysis was a viable option to enzymatic hydrolysis. It was observed that, at 70 °C, quick conversion of lignocellulose biomass to monosaccharides takes place, and therefore high sugar recovery efficiency is achieved. The economic viability of using H_2SO_4 was examined by Goldstein and Easter (1992), Goldstein et al. (1989), and Yu et al. (2008). They concluded that to recover and condense the acid by electrodialysis, while ensuring minimization of the acid to substrate ratio and also the capability of reusing it for further hydrolysis, makes it economically attractive. It must be noted that a higher acid- to-substrate ratio would negatively affect fermentation.

Sun et al. (2011) presented a viable approach based on using sulfuric- acid hydrolysis (Fig. 4.13) to produce fuel ethanol from bamboo biomass. During the process, colour compounds were removed with activated carbon, while separation and recovery of the acid and sugar from the saccharified liquid was achieved using an improved simulated moving bed (ISMB) amounting to 90.5% (for acid recovery) and 98.4% (for sugar recovery). After this, continuous hydrolysis of the oligosaccharides within the sugar fraction coupled with pH adjustments was accomplished. Finally, the unsterilised saccharified solution was made to undergo continuous fermentation using the flocculating-yeast (saccharomyces cerevisiae) strain for KF-7 bamboo ethanol. This process showed a sugar recovery efficiency of 81.6%, fermentation yield of 92.0%, ethanol concentration of 27.2 g/l, and a high ethanol productivity value of 8.2 g/l/h. A general flow process chart of the production of ethanol from bamboo is shown in Fig. 4.14.

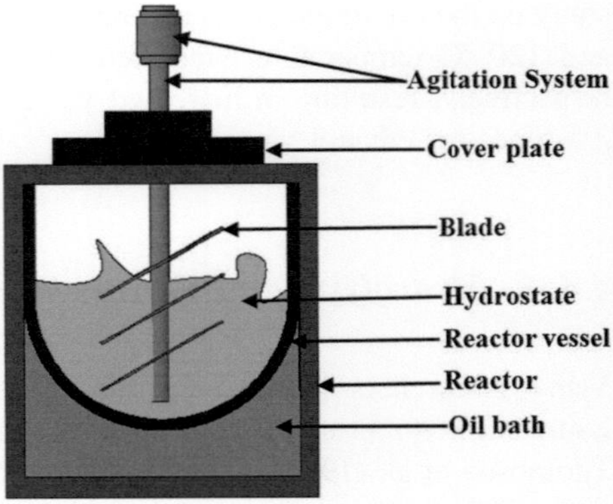

Fig. 4.13 Reactor system used for concentrated sulfuric-acid hydrolysis

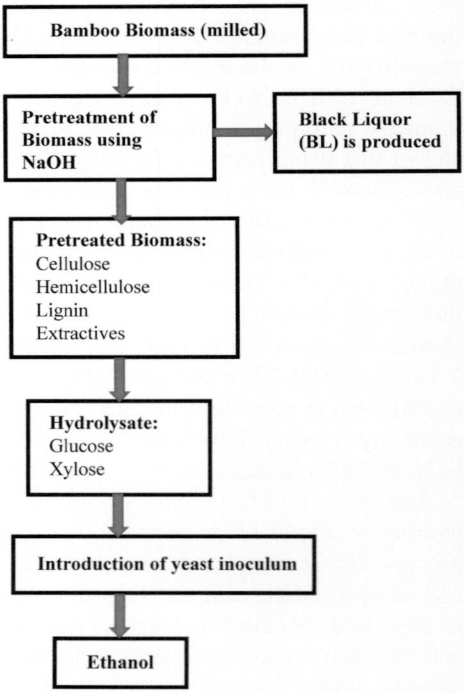

Fig. 4.14 Flow process of ethanol production from bamboo. *Adapted from* Kuttiraja et al. (2013)

4.4.5 *Use of Immature Bamboo Shoots*

Researchers, such as Demenezes et al. (1983) and Ram and Seenayya (1991), worked on improving several pre-treatment methods for increasing the rate of enzymatic saccharification for the production of bioethanol from mature bamboo. However, the lignin content in mature bamboo is higher than for immature bamboo shoots (Fujii et al. 1993). This implies that much more sophisticated and expensive pre-treatment processes may be required when dealing with the industrial production of ethanol if mature bamboo is used. Shimokawa et al. (2009) experimented with the use of immature bamboo shoots for the production of ethanol. During the research, it was observed that lignin was uniformly distributed and spread out within the culm once branching has been initiated in bamboo. Immature bamboo shoots produce lower levels of lignin that cannot inhibit enzymatic hydrolysis. Another advantage is that pre-treatment is not a requirement for achieving high saccharification yield. The results showed that, during enzymatic hydrolysis using xylanase and cellulase, there was an increase of about 98% in saccharification yield for immature *Phyllostachys bambusoides*. The simultaneous saccharification and fermentation, using the enzyme 12 FPU gl for hydrolysis and the yeast Saccharomyces cerevisiae for fermentation, resulted in 81 and 98% ethanol yields for *Phyllostachys pubescens* and *bambusoides*, respectively.

4.4.6 Summary

In conclusion, the production of bioethanol from bamboo is a viable energy resource that is renewable and sustainable. While both mature and immature shoots of bamboo can be utilised for ethanol production, however, mature bamboo lignocellulose requires the use of pre-treatment and acid hydrolysis, immature bamboo shoots do not need pre-treatment, and enzymatic hydrolysis can be used. Bamboo represents a non-food bioethanol fuel which can provide an alternative to fossil fuel in many developing countries in Asia and Africa.

4.5 Production of Methane from Bamboo

Anaerobic digestion, as a means of obtaining renewable energy from lignocellulose biomass such as bamboo, is receiving increased interest because of global warming fears, fossil fuel depletion, and little competing demands from human and animals as food (Crespo et al. 2012). Fengel and Wengener (1983) described a biomass material as a composite consisting of cellulose reinforcement and a matrix of hemicellulose and lignin. In anaerobic digestion, organic-matter decomposition is caused by microorganisms, in the absence of oxygen, to produce methane biogas, composed of 50–75% CH_4 and 25–50% CO_2 (Frigon and Guiot 2010). Deublein and Steinhauser (2011) reported that the anaerobic-digestion methodology and technology for production of methane biogas are much more efficient than the production of ethanol. Rutz and Janssen (2008) concluded that anaerobic digestion had some key environmental benefits, such as reducing methane emission which

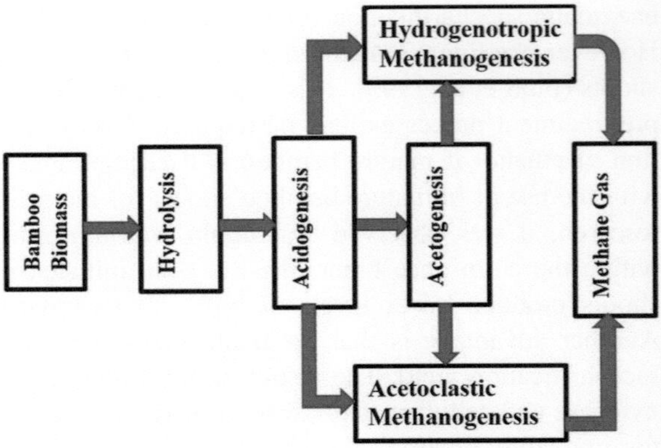

Fig. 4.15 Anaerobic digestion flow process. *Adapted from* (Li et al. 2011)

has 20 times the global-warming impact of carbon dioxide (CO_2) of landfills and other open-air environments. Bamboo is a valuable biomass that is a rich source of renewable energy because it is composed of cellulose and hemicellulose which are known as holocellulose. Holocellulose is essential for the production of methane in plant biomass. However, a strong network of lignin encapsulates it. Bamboo is a lignocellulose made up of 57–65% cellulose, 27–30% hemicelluloses, and 4.9–5.0% lignin (Asada et al. 2005; Shimokawa et al. 2009; Yamashita et al. 2010; Sun et al. 2011, 2013; Arisutha et al. 2016).

The four main steps for anaerobic digestion as described in Fig. 4.15 are summarised as follows:

1. Hydrolysis

   The complex organic polymer is decomposed into soluble monomers by microbes through enzymatic hydrolysis. Further hydrolysis continues as protein is hydrolysed to amino acids, lipids to fatty acids, and carbohydrates to sugars.

2. Fermentation

   The fermentation of methane using microorganisms can be described by three major processes, namely: saccharification (hydrolysis), production of acid, and production of gas (Fukuma et al. 1993). Fermentative bacteria, also known as acidogens, convert the hydrolysed substrates to volatile fatty acids (VFAs), alcohols, etc.

3. Acetogenic bacteria then act on the VFAs mixture to produce acetic acid, carbon dioxide (CO_2), and hydrogen.
4. The substrate from acetogenesis undergoes methanogenesis to produce methane.

The most critical stage is methanogenesis because methanogens characteristically have the slowest rate of growth and are therefore most sensitive to the slightest alteration in pH, environment, temperature, and concentration levels of inhibitors. Methanogenesis is therefore termed a rate-limiting step in anaerobic digestion

(Chen et al. 2008). The second rate-limiting step is hydrolysis of lignocellulose biomass (Cirne et al. 2007). The rate-limiting steps in the anaerobic digestion (AD) process make the introduction of pre-treatment preceding AD extremely crucial for improving methane yield in the production of biomass.

4.5.1 *Resistance to Hydrolysis*

The resistance to enzymatic attack is mainly a function of the lignin content the biomass has within its cell walls. Lignin content directly affects the saccharification by limiting the sugar conversion ratio of cellulose and hemicellulose (Wayman and Obiaga 1974; Azuma and Koshijima 1983; Nakamura et al. 1997). Klasson lignin is an inhibitor to methane production because of its capability to prevent microorganisms initiating saccharification.

4.5.2 *Pre-treatment*

To break or degrade the lignin, pre-treatment is required. The section on ethanol production has summarised some of the pre-treatments that are suitable for ensuring enzymatic hydrolysis. Pre-treatments may be classified as physical, thermal, chemical, thermos-chemical, or biological (Kuwahara et al.1984; Scurlock et al. 2000; Kurakake et al. 2001; Kobayashi et al. 2004; Wyman et al. 2005; Mladenovska et al. 2006; Wang and Ren 2008; Fernandes et al. 2009; Xie et al. 2011; Zhong et al. 2011; Zilin et al. 2012; Frigon et al. 2012; Sun et al. 2013). The main purpose of introducing pre-treatment methods is to cause the destruction or partial destruction of hemicellulose and lignin, increase material porosity, and reduction in crystallinity of cellulose. An efficient and effective pre-treatment process is intended to accomplish the following:

- Enhance the sugar formation that leads to improved enzymatic hydrolysis
- Reduce carbohydrate loss or degradation
- Inhibit formation of toxic by-products during anaerobic digestion by bacteria (Sun et al. 2002).

Pre-treatment is implemented to facilitate sugar release, such as monomeric hexose and pentose from enzymatic hydrolysis of cellulose and hemicellulose and, subject to anaerobic consortium, to produce methane. The pre-treatment process introduces hemicellulose and ensures better accessibility to biochemical and chemical degradation. During hydrolysis, polymer bonds are broken to release xylose which degrades further to furfural (5-hydroxymethylfurfura) (Fig. 4.16). Saccharification is, therefore, the process whereby polysaccharides undergo hydrolysis to release soluble sugars.

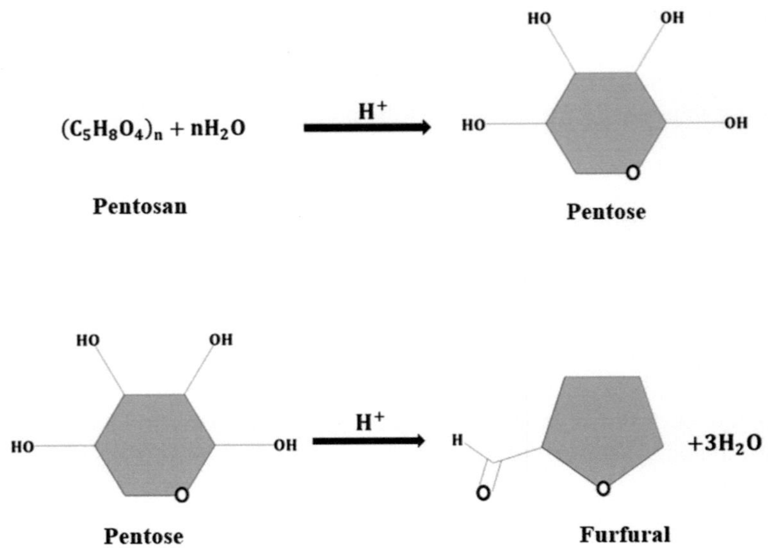

Fig. 4.16 Hydrolysis of hemicellulose to xylose and then to furfural

4.5.2.1 Steam-Explosion Pre-treatment

One of the more effective pre-treatment methods that have been successfully utilised for enzymatic saccharification and to ensure fermentation and improve alcohol yields is the steam-explosion process. Mason et al. (1926) first designed and patented the steam-explosion pre-treatment process for wood. The main components of the steam-explosion equipment are; a steam generator, high-pressure reactor, a receiver, and a condenser equipped with a silencer (Nakamura et al. 1989, 1991; Kobayashi et al. 1998; Sawada et al. 1995). The steam-explosion process can be defined as treating biomass with hot steam at temperatures between 180 and 240 °C and under pressures between 1 and 3.5 Mpa, inducing a rupture of fibres within the biomass as a direct result of an explosive decomposition. Defibrillation of cellulose bundles occurs during the sudden pressure release, ensuring access to enzymatic hydrolysis and fermentation. Garrote et al. (1999) reported the following advantages of the steam-explosion pre-treatment process for lignocellulose:

- Only water is used, and it is therefore chemical free
- Less corrosive to equipment in comparison to pre-treatment processes such as acid hydrolysis
- Yield of hemicellulose is very good
- The handling and recycling stages of using acids for pre-treatment are unnecessary.

Steam explosion can, therefore, be considered as an environmentally friendly process for biogas and ethanol production from biomass (Kokta and Ahmed 1998).

Alternative techniques, e.g. steam refining, have been utilised by researchers such as Schütt et al. (2012) who compared it with conventional steam explosion and found no significant difference in enzymatic hydrolysis activities between the two processes. Kobayashi et al. (2004) investigated the use of steam-explosion pre-treatment to produce methane from bamboo. The study revealed that 1gr of bamboo produced 215 ml of methane gas when steam at a pressure of 3.5 Mpa is used in the pre-treatment of bamboo. An inverse relationship between the quantity of methane production and lignin levels was observed. Pre-treatment is therefore critical to methane production because it degrades Klason lignin, which is the primary inhibitory to enzymatic saccharification. Kobayashi et al. (2004) carried out cultures on digestion sludge for use as original microbial seed for the enzymatic saccharification and fermentation processes. The biogas produced was harvested using water displacement with a solution of saturated sodium.

4.5.2.2 Chemical Pre-treatments

Shen et al. (2014) conducted a study that showed the relationship between four pre-treatment methods (acid, alkaline, alkaline added enzyme; and enzyme) and the chemical oxygen demand, sugar yield; and methane yield. Dewatered sewage sludge was used to feed a mesophilic biogas digester to produce anaerobic sludge which was used as inoculums (pH: 7.6, partial alkalinity: 6.5 g/l). It was observed that: pre-treatment is key to improving COD solubilisation; alkaline-aided enzyme pre-treatment gave the highest sugar yield (but not the highest methane yield): and the best pre-treatment method that gave the highest methane yield (88%) was alkaline pre-treatment. All pre-treatments used led to an enhanced methane production rate in comparison with using untreated bamboo. The conclusion can, therefore, be made that alkaline pre-treatment is a viable and cost-efficient way of producing methane from bamboo because alkaline materials boost anaerobic digestion. The major drawback of chemical pre-treatment is the relatively high cost of the chemicals.

4.5.2.3 Thermo-Chemical Pre-treatments

Arisutha et al. (2016) investigated the effects of thermochemical pre-treatment methods, including calcium hydroxide ($Ca(OH)_2$), sodium hydroxide (NaOH), ammonium carbonate ($(NH_4)_2CO_3$), sulphuric acid (H_2SO_4), maleic acid ($HO_2CCHCHCO_2H$), and hydrothermal with H_2SO_4 in the production of bamboo biogas. The pre-treatment was carried out in an anaerobic digestor. Results from the above scenarios, in descending order, show hydrothermal with H_2SO_4, as producing the highest methane yield (72%), and then H_2SO_4, $Ca(OH)_2$, maleic acid, and $(NH_4)_2CO_3$. The hydrothermal with H_2SO_4 can be an effective pre-treatment process that can enhance the methane yield in bamboo biogas. The major drawback of chemical pre-treatment is the relatively high cost of the chemicals.

4.5.2.4 Biological Pre-treatment

The main biological pre-treatment methods involve the use of fungus, microbial consortium, and enzymes. In comparison with other pre-treatment methods, such as physical and chemical processes, fewer inhibitors are generated, no chemicals are required, and the energy requirement is far lower. However, the disadvantages include microbial competition for carbohydrate downstream and the long pre-treatment time, which is a limiting factor in its application on a commercial level. The overall objective of biological pre-treatment is the minimisation of carbohydrate loss and maximisation of lignin removal. Most biological pre-treatments are less effective and efficient in comparison to chemical treatments. A combination of the two pre-treatments has been shown to improve methane yields significantly. Michalska et al. (2013) observed that, when NaOH treatment and enzymatic hydrolysis were used in combination, methane yields of about 300% were achieved.

4.5.2.5 Fungal Pre-treatment

Fungal pre-treatment is conducted under very sterile conditions using several classes of fungi such as soft-rot, brown-rot, and white-rot fungi to perform selective degradation of hemicellulose and lignin while having a minimum effect on cellulose. Cellulose is naturally more resistant to fungal attacks, and white-rot fungus is the most effective during pre-treatment (Sun and Cheng 2002).

4.5.2.6 Microbial Consortium Pre-treatment

Microbial consortium pre-treatment refers to the use of screened microbes on lignocellulose biomass. In comparison with fungal pre-treatment, the sterilisation of lignocellulose biomass is not a requirement. However, the microbes also degrade cellulose and hemicellulose and are therefore not lignin selective (Zhang et al. 2014).

4.5.2.7 Enzymatic Pre-treatment

Enzymatic pre-treatment involves the use of enzymes such as cellulase and hemicellulase in the hydrolysis of cellulose and hemicellulose in lignocellulose biomass. The methane yield from enzymatic pre-treatment is minimal, coupled with a high enzyme cost, which limits its commercial appeal.

4.5.2.8 Summary

From the various literature reviewed, it is proven that bamboo may be a viable biomass for the production of methane gas for energy use. Both old and new sprouting shoots can be hydrolyzed to produce methane gas, and the most critical stage is the pre-treatment stage.

References

Adapa PK, Singh AK, Schoenau GJ, Tabil LG (2006) Pelleting characteristics of fractionated alfalfa grinds: hardness models. Power Handling and Process 18:1–6

Afrin T, Tsuzuki T, Wang X (2009) Bamboo fibres and their unique properties, Natural fibres in Australasia. In: proceedings of the combined (NZ and AUS) Conference of The Textile Institute, Dunedin, New Zealand, 15–17 Apr 2009, pp 77–82

Alizadeh H, Teymouri F, Gilbert TI, Dale BE (2005) Pretreatment of switchgrass by ammonia fiber explosion. Appl Biochem Biotechnol 1133–1141

Anyang Gemco Energy Machinery (2016) 7 steps of fuel pellets production process. http://www.pelletmillsolution.com/blogs/7-Steps-of-Fuel-Pellets-Production-Process.html. Accessed 23 Sept 2016

Arisutha S, Baredar P, Deshpande DM, Suresh S (2016) Effects of thermo-chemical pre-treatment on bamboo for biogas production. Indian Institute of Chem Eng 58(1):79–88

Asada C, Nakamura Y, Kobayashi F (2005) Waste reduction system for production of useful materials from un-utilized bamboo using steam explosion followed by various conversion methods. Biochem Eng J 23:131–137

Azuma J, Koshijima T (1983) Hydrolysis of wood polysaccharides by enzyme system. Mokuzai Kenkyu Shiryo 17:1–20

Beall FC, Eickner HW (1970) Thermal degradation of wood components: a review of the Literature. U.S.D.A. Forest service, research paper FPL 130

Binder JB, Raines RT (2010) Fermentable sugars by chemical hydrolysis of biomass. PNAS 107:4516–4521

Brownell HH, Yu EKC, Saddler JN (1986) Steam-explosion pretreatment of wood: effect of chip size, acid, moisture content and pressure drop. Biotechnol Bioeng 28:792–801

Chang VS, Holtzapple MT (2000) Fundamental factors affecting enzymatic reactivity. Appl Biochem Biotechnol 5–37

Chang VS, Kaar WE, Burr B, Holtzapple MT (2001) Simultaneous saccharification and fermentation of lime-treated biomass. Biotechnol Lett 23(16):1327–1333

Chang VS, Nagwani M, Holtzapple MT (1998) Lime pretreatment of crop residues bagasse and wheat straw. Appl Biochem Biotechnol 74:135–159

Chen Y, Cheng JJ, Creamer KS (2008) Inhibition of anaerobic digestion process: a review Bioresour Technol 99: 4044–4064

Cirne DG, Lehtomäki A, Björnsson L, Blackall LL, (2007) Hydrolysis and microbial community analyses in two-stage anaerobic digestion of energy crops. J Appl Microbiol 103(3):516–527

Clausen EC, Gaddy JL (1993) Concentrated sulfuric acid process for converting lignocellulosic materials to sugars. US patent, No. 5, 188, 673

Cowling EB, Kirk TK (1976) Properties of cellulose and lignocellulosic materials as substrates for enzymatic conversion process. Biotechnol Bioeng Symp 6:95–123

Crespo CF, Badshah M, Alvarez MT, Mattiasson B (2012) Ethanol production by continuous fermentation of d-(+)-cellobiose, d-(+)-xylose and sugarcane bagasse hydrolysate using the thermoanaerobe caloramator boliviensis. Bioresour Technol 103:186

Delgenés JP, Penaud V, Moletta R (2002) Pretreatments for the enhancement of anaerobic digestion of solid wastes Chapter 8. In: Biomethanization of the organic fraction of municipal solid wastes. IWA Publishing, pp 201–228
Demenezes TJB, Dossantos CLM, Azzini A (1983) Saccharification of bamboo carbohydrates for the production of ethanol. Biotechnol Bioeng 25:1071–1082
Deublein D, Steinhauser A (2011) Biogas from Waste and Renewable Resources. 2nd edn. Wiley-VCH Verlag, Weinheim
Domansky R, Rendos F (1962) On the pyrolysis of wood and its components. Holz Roh Werkst 29:473–476
FAO (2009) State of the World's forests. Food and Agriculture Organisation of the UN, Rome, Italy
FAO (2013) Statistical yearbook 2013. http://www.fao.org/docrep/018/i3107e/i3107e.PDF. World Food and Agriculture. Accessed Oct 2016
Food and Agriculture Organization of the United Nations (FAO) (2010). Criteria and indicators for sustainable wood fuels. Rome
Faizal HM, Latiff ZA, Wahid MA, Darus AN (2010) Physical and combustion characteristics of biomass residues from palm oil mills. In: New aspects of fluid mechanics, heat transfer and environment, pp 34–38.
Fan LT, Gharpuray MM, Lee YH (1987) Cellulose hydrolysis. Springer-Verlag, Berlin
Farrell AE, Plevin RJ, Turner BT, Jones AD, O'Hare M, Kammen DM (2006) Ethanol can contribute to energy and environmental goals. Science 311:506–508
Fengel D, Wegener G (1983) Wood: Chemistry, Ultrastructure, Reactions. Walter de Gruyter, Berlin
Fernandes TV, Klaasse Bos GJ, Zeeman G, Sanders JPM, Van Lier JB (2009) Effect of thermo-chemical pre-treatment on anaerobic biodegradability and hydrolysis of lingo-cellulosic biomass. Bioresour Technol 100:2575–2579
Frigon J-C, Guiot SR (2010) Biomethane production from starch and lignocellulosic crops: a comparative review. Biofuels, Bioproducts and Biorefining, 4(4):447–458
Frigon JC, Mehta P, Guiot SR (2012) Impact of mechanical, chemical and enzymatic pretreatments on the methane yield from the anaerobic digestion of switchgrass. Biomass Bioenergy 36:1–11
Fu DB, Mazza G (2011) Aqueous ionic liquid pretreatment of straw. Bioresour Technol 102:7008–7011
Fujii Y, Azuma J, Marchessault RH, Morin FG, Aibara S, Okamura K (1993) Chemical composition change of bamboo accompanying its growth. Holzforschung 47:109–115
Fukuma M, Takesada K, Yasunishi A (1993) Methane production through two-phase anaerobic treatment of wastewater containing cellulose with high solid content and some easily digestible organic matters. Kagaku Kogaku Ronbunsyu 19:885–892
Fu JH, Chen W (2009) Profile of bamboo charcoal museum in Suichang County, Zhejiang 2009. World Bamb Rattan 7(5):1–45
Gandi J, Holtzapple MT, Ferrer A, Byers FM, Turner ND, Nagwani M, Chang S (1997) Lime treatment of agricultural residues to improve rumen digestibility. Anim Feed Sci Technol 68:195–211
Garrote G, Dominguez H, Parajo JC (1999) Hydrothermal processing of lignocellulosic materials. Holzforschung 57:191–202
Goldstein IS, Bayat-Makooi F, Sabharwal HS, Singh TM (1989) Acid recovery by electrodialysis and its economic implications for concentrated acid hydrolysis of wood. Appl Biochem Biotechnol 20(21):95–106
Goldstein IS, Easter JM (1992) An improved process for converting cellulose to ethanol. Tappi J 8:135–140
Gossett JM, Stuckey DC, Owen WF, McCarty PL (1982) Heat treatment and anaerobic digestion of refuse. J Environ Eng Div 108:437–454
Gray KA, Zhao LS, Emptage M (2006) Bioethanol. Curr Opin Chem Biol 10:141–146

Gregg D, Saddler JN (1996) A techno-economic assessment of the pretreatment and fractionation steps of a biomass-to-ethanol process. Appl Biochem Biotechnol 711–727

Grethlein HE (1985) The effect of pore size distribution on the rate of enzymatic hydrolysis of cellulose substrates. Bio/Technol 3:155–160

Grous WR, Converse AO, Grethlein HE (1986) Effect of steam explosion pretreatment on pore size and enzymatic hydrolysis of poplar. Enzyme Microbiol Technol 8:274–280

Hartmann H, Angelidaki I, Ahring BK (1999) Increase of anaerobic degradation of particulate organic matter in full-scale biogas plants by mechanical maceration. In: Mata-Alvarez J, Tilche A, Cecchi F (eds) Proceedings of the second international symposium on anaerobic digestion of solid wastes, Barcelona, vol 1, pp 129–136

Hendriks ATWM, Zeeman G (2009) Pretreatments to enhance the digestibility of lignocellulosic biomass. Bioresour Technol 100:10–18

Hon DNS, Shiraishi N (2001) Wood and cellulosic chemistry, 2nd edn. Deker, New York

Iranmahboob J, Nadim F, Monemi S (2002) Optimizing acid-hydrolysis: a critical step for production of ethanol from mixed wood chips. Biomass Bioenergy 22:401–404

Kim S, Dale BE (2004) Global potential bioethanol production from wasted crops and crop residues. Biomass Bioenerg 26:361–375

Kirk-Othmer (2001) Encyclopedia of chemical technology. Concise. 4th edn. Wiley Interscience

Kobayashi F, Sawada T, Nakamura Y, Ohnaga M, Godliving M, Ushiyama T (1998) Saccharification and alcohol fermentation in starch solution of steam-exploded potato. Appl Biochem Biotechnol 69:177–189

Kobayashi F, Take H, Asada C, Nakamura Y (2004) Methane production from steam-exploded bamboo. J Biosci Bioeng 97:426–428

Kohlmann KL, Westgate PJ, Sarikaya A, Velayudhan A, Weil J, Hendrickson R et al (1995) Enhanced enzyme activities on hydrated lignocellulosic substrates. BTEC paper 127. In: 207th American Chemical Society National meeting, ACS Symposium series No. 618. Enzymatic degradation of insoluble carbohydrates, pp 237–255

Kokta BV, Ahmed A (1998) Chapter 6, steam explosion pulping, environmentally friendly technologies for the Pulp and Paper Industry. Wiley, Hoboken.

Koullas DP, Christakopoulos P, Kekos D, Macris BJ, Koukios EG (1992) Correlating the effect of pretreatment on the enzymatic hydrolysis of straw. Biotechnol Bioeng 39:113–116

Kumar R, Chandrashekar N (2014) Fuel properties and combustion characteristics of some promising bamboo species in India. Northeast Forestry University and Springer-Verlag, Berlin Heidelberg

Kurakake M, Kisaka W, Ouchi K, Komaki T (2001) Pretreatment with ammonia water for enzymatic hydrolysis of corn husk, bagasse, and switchgrass. Appl Biochem Biotechnol 90:251–259

Kuttiraja M, Sindhu R, Varghese PE, Sandhya SV, Binod P, Vani S et al (2013) Bioethanol production from bamboo (Dendrocalamus sp.) process waste, biomass and bioenergy, vol 59. Elsevier, Amsterdam, pp 142–150.

Kuwahara M, Kagimura T, Takagi K (1984) Anaerobic Fermentation of Bark I. Effect of Treatment of Bark with White-rot Fungi and Chemicals on the Production of methane. Mokuzai Gakkaishi 30:769–776

Laser M, Schulman D, Allen SG, Lichwa J, Antal MJ Jr, Lynd LR (2002) A comparison of liquid hot water and steam pretreatments of sugar cane bagasse for bioconversion to ethanol. Bioresour Technol 81:33–44

Laureano-Perez L, Teymouri F, Alizadeh H, Dale BE (2005) Understanding factors that limit enzymatic hydrolysis of biomass. Appl Biochem Biotechnol 1081–1099

Li CL, Cheng G, Balan V, Kent MS, Ong M, Chundawat SPS, Sousa L, Melnichenko YB, Dale BE, Simmons BA, Singh S (2011) Influence of physico-chemical changes on enzymatic digestibility of ionic liquid and AFEX pretreated corn stover. Bioresour Technol 102: 6928–6936

Liu Z, Jiang Z, Cai Z, Fei B, Yu Y, Liu X (2012) The Manufacturing Process of Bamboo Pellets. In: Proceedings of the 55th International Convention of Society of Wood Science and Technology, Beijing, 27–31 August 2012

Liu C, Wyman CE (2003) The effect of flow rate of compressed hot water on xylan, lignin and total mass removal from corn stover. Ind Eng Chem Res 42:5409–5416

Magellia F, Boucherb K, Bib HT, Melinc S, Bonolia A (2009) An environmental impact assessment of exported wood pellets from Canada to Europe, biomass and bioenergy, vol 33, pp 434–441

Malisius U, Jauschnegg H, Schmidt H, Nilsson B, Rapp S, Strehler H (2000) Wood pellets in Europe. Industrial network on wood pellets. UMBERA GmbH, A-3100 St. Poelten, Schiessstattring

Mason WH (1926) Process and apparatus for disintegration of wood and the like. US Patent: 1578609

Michalska K, Ledakowicz S (2013) Alkali pre-treatment of Sorghum Moench for biogas production. Chem Pap 67:1130–1137

Mladenovska Z, Hartmann H, Kvist T, Sales-Cruz M, Gani R, Ahring BK (2006) Thermal pretreatment of the solid fraction of manure: Impact on the biogas reactor performance and microbial community. Water Sci Technol 53:59–67

Mosier N, Hendrickson R, Ho N, Sedlak M, Ladisch MR (2005a) Optimization of pH controlled liquid hot water pretreatment of corn stover. Bioresour Technol 96:1986–1993

Mosier N, Wyman C, Dale B, Elander R, Lee YY, Holtzapple M, Ladisch M (2005b) Features of promising technologies for pretreatment of lignocellulosic biomass. Bioresour Technol 96:673–686

Nakamura Y, Mohammed M, Nagao M, Sawada T, Motoi M (1991) Modification of rice straw by steam explosion and enzymatic saccharification of steam-exploded products. Kagaku Kogaku Ronbunsyu 17:504–510

Nakamura Y, Sawada T, Kuwahara M, Motoi M (1989) Effect of explosion on saccharification and fermentation of biomass. Mokuzai Gakkaishi 35:663–668

Nakamura Y, Sawada T, Sungusia MG, Kobayashi F, Kuwahara M, Ito H (1997) Lignin peroxidase production by Phanerochaete chrisosporium immobilized on polyurethane foam. J Chem Eng Jpn 30:1–6

Palmowski L, Muller J (1999) Influence of the size reduction of organic waste on their anaerobic digestion. In: II International symposium on anaerobic digestion of solid waste. Barcelona, 15–17 June, pp 137–144

Palonen H, Thomsen AB, Tenkanen M, Schmidt AS, Viikari L (2004) Evaluation of wet oxidation pretreatment for enzymatic hydrolysis of softwood. Appl Biochem Biotechnol 117:1–17

Pilusa TJ, Huberts R, Muzenda E (2013) Emissions analysis from combustion of eco-fuel briquettes for domestic applications. J Energy in Southern Afr 24(4)

Puri VP (1984) Effect of crystallinity and degree of polymerization of cellulose on enzymatic saccharification. Biotechnol Bioeng 26:1219–1222

Ram MS, Seenayya G (1991) Production of ethanol from straw and bamboo pulp by primary isolates of Clostridium thermocellum. World J Microb Biotechnol 7:372–378

Ramos LP (2003) The chemistry involved in the steam treatment of lignocellulosic materials. Quim Nova 26(6):863–871

Renewable Energy Policy Network for the 21st Century (2016) Renewables 2016 Global status report. http://Www.Ren21.Net/Wp-Content/Uploads/2016/06/Gsr_2016_Full_Report.Pdf. Accessed 17 Sept 2016

Rubin EM (2008) Genomics of cellulosic biofuels. Nature 454:841–845

Rutz D, Janssen R (2008) Biofuel Technology Handbook. WIP Renewable Energies, München

Saha BC (2003) Hemicellulose bioconversion. J Ind Microbiol Biotechnol 30:279–291

Sander KB, Hyseni, Haider W (2011) Wood-based biomass energy development for Sub-Saharan Africa: issues and approaches, ed. A. E. Unit. 2011. The World Bank, Washington, DC

Sathitsuksanoh N, Zhu ZG, Ho TJ, Bai MD, Zhang YHP (2010) Bamboo saccharification through cellulose solvent-based biomass pretreatment followed by enzymatic hydrolysis at ultra-low cellulase loadings. Bioresour Technol 101:4926–4929

Sawada T, Nakamura Y, Kobayashi F, Kuwahara M, Watanabe T (1995) Effects of fungal pretreatment and steam explosion pretreatment on enzymatic saccharification of plant biomass. Biotechnol Bioeng 48:719–724

Schütt F, Westereng B, Horn SJ, Puls J, Saake B (2012) Steam refining and alternative steam explosion. Biores Technol 111:476–481

Scurlock JMO, Dayton DC, Hames B (2000) Bamboo: an overlooked biomass resource. Biomass Bioenergy 19:229–244

Shen S, Nges IA, Yun J, Liu J (2014) Pre-treatments for enhanced biochemical methane potential of bamboo waste. Chem Eng J 240:253–259

Shimokawa T, Ishida M, Yoshida S, Nojiri M (2009) Effects of growth stage on enzymatic saccharification and simultaneous saccharification and fermentation of bamboo shoots for bioethanol production. Bioresour Technol 100:6651–6654

Sims REH, Mabee W, Saddler JN, Taylor M (2010) An overview of second generation biofuel technologies. Bioresour Technol 101:1570–1580

Song Z, Yang G, Guo Y, Zhang T (2012) Comparison of two chemical pretreatments of rice straw for biogas production by anaerobic digestion. BioResource 7(3):3223–3236

Steinfeld C (2001) A bamboo future, *Environmental Design and Construction*, 1–5

Sticklen MB (2008) Plant genetic engineering for biofuel production: towards affordable cellulosic ethanol. Nat Rev Genet 9:433–443

Sun Y, Cheng J (2002) Hydrolysis of lignocellulosic materials for ethanol production: a review. Bioresour Technol 83

Sun Y, Lin L (2010) Hydrolysis behavior of bamboo fiber in formic acid reaction system. J Agric Food Chem 58:2253–2259

Sun ZY, Tang YQ, Iwanaga T, Sho T, Kida K (2011) Production of fuel ethanol from bamboo by concentrated sulfuric acid hydrolysis followed by continuous ethanol fermentation. Bioresour Technol 102:10929–10935

Sun ZY, Tang YQ, Morimura S, Kida K (2013) Reduction in environmental impact of sulfuric acid hydrolysis of bamboo for production of fuel ethanol. Bioresour Technol 128:87–93

Teixeira LC, Linden JC, Schroeder HA (1999) Alkaline and peracetic acid pretreatments of biomass for ethanol production. Appl Biochem Biotechnol 19–34

Telmo C, Lousada J (2011) Heating values of wood pellets from different species. Biomass Bioenerg 35:2539–2634

Thompson DN, Chen HC, Grethlein HE (1992) Comparison of pretreatment methods on the basis of available surface area. Bioresour Technol 39:155–163

UNDP, WHO (2009) The energy access situation in developing countries: a review focusing on the least developed countries and Sub-Saharan Africa. In: Legros G et al (eds). New York, USA

Vinterback J (2004) Pellets 2002: the first world conference on pellets. Biomass Bioenerg 27: 513–520

Wang X, Ren H (2008) Comparative Study of the Photo-discoloration of Moso Bamboo (Phyllostachys pubescens Mazel) and Two Wood Species. Appl Surf Sci 254:7029–7034

Wayman M, Obiaga TI (1974) The modular structure of lignin. Can J Chem 52:2102–2110

Weil JR, Brewer M, Hendrickson R, Sarikaya A, Ladisch MR (1997) Continuous pH monitoring during pretreatment of yellow poplar wood sawdust by pressure cooking in water. Appl Biochem Biotechnol 68:21–40

Weil JR, Sarikaya A, Rau SL, Goebz J, Lasisch CM, Brewer M, Hendrickson R, Tadisch MR (1998) Pretreatment of corn fiber by pressure cooking in water. Appl Biochem Biotechnol 73:1–17

Wen JL, Xiao LP, Sun YC, Sun SN, Xu F, Sun RC, Zhang XL (2011) Comparative study of alkali-soluble hemicelluloses isolated from bamboo (Bambusarigida). Carbohydr Res 346: 111–120

Wyman CE, Dale BE, Elander RT, Holtzapple M, Ladisch MR, Lee YY (2005) Comparative sugar recovery data from laboratory scale application of leading pretreatment technologies to corn stover. Bioresour Technol 96:2026–2032
Xie S, Frost JP, Lawlor PG, Wu G, Zhan X (2011) Effects of thermo-chemical pretreatment of grass silage on methane production by anaerobic digestion. Bioresour Technol 102:8748–8755
Xu F (2010) Structure, ultrastructure, and chemical composition. In: Sun RC (ed) Cereal straw as a resource for sustainable biomaterials and biofuels. Elsevier, Amsterdam, pp 9–47
Yamashita Y, Shono M, Sasaki C, Nakamura Y (2010) Alkaline peroxide pretreatment for efficient enzymatic saccharification of bamboo. Carbohyd Polym 79:914–920
Yu WK, Chung KF, Chan SL (2003) Column buckling of structural bamboo, engineering structures, vol 25, no 6. Elsevier, Amsterdam, 755–768
Yu Y, Lou X, Wu HW (2008) Some recent advances in hydrolysis of biomass in hot-compressed, water and its comparisons with other hydrolysis methods. Energy Fuels 22:46–60
Zhang XY, Yu HB, Huang HY, Liu YX (2007) Evaluation of biological pretreatment with white rot fungi for the enzymatic hydrolysis of bamboo culms. Int Biodeter Biodegr 60:159–164
Zhong W, Zhongzhi Z, Wei Q (2011) Comparison of Chemical and Biological Preteatment of Corn Straw for Biogas Production by Anaerobic Digestion. Renew Energ 36(6):1875–1879
Zhang L, You T, Zhang L, Li M, Xu F (2014) Comprehensive utilization of waste hemicelluloses during ethanol production to increase lactic acid yield: from pretreatment to fermentation. Biotechnology for Biofuels, 7:494 doi: 10.1186/s13068-014-0189-4

Chapter 5
Applications of Bamboo

Abstract The general applications of bamboo, in recent years, cuts across the fields of engineering, as well as non-engineering fields. This chapter offers a thorough review of some traditional and non-traditional uses of bamboo to demonstrate that it is indeed a multipurpose plant. For many centuries, it has been primarily the construction industry that has benefited from using bamboo. In recent times, however, the trend has evolved to one in which other fields such as the textile, pulp and paper, medical, arts and design, aerospace, and food and beverage industries are researching and utilising all facets of the bamboo plant. Its widespread use has led to many research opportunities and projects that are critically investigating all properties and extractives useful and applicable to even more recently emerging fields, such as nanotechnology. The review shows that all parts of the bamboo plant can be harnessed and utilised. The economic importance of this plant to most developing countries in Africa, Asia, and South America will continue to grow.

5.1 Introduction

Bamboo is one of the oldest structural material used by humankind (Latif and Liese 1995). It has been used widely for household products and extended to industrial applications due to technological advancement, resulting in increased market demand for it as a raw material and for its products For many centuries, bamboo culms have been used for constructing all kinds of structures and structural components, such as houses, shelters, boards, roof trusses, wall cladding, pillars, columns, tools, skyscraper scaffolding, furniture, flooring, ceilings, walls, windows, doors, fences, rafters, and purlins. It can also be used for soil stabilisation, rehabilitation of degraded lands, medicine, handicrafts, biomass, fuel, food (the shoots), water transportation facilities, paper and pulp, and many other uses (Van der Lugt et al. 2005; Yu et al. 2011). Bamboo is also used for household utilities such as containers, chopsticks, woven mats, fishing poles, cricket boxes, handicrafts, and chairs. The culms are used as load-bearing elements in construction, such as in

E.T. Akinlabi et al., *Bamboo*, DOI 10.1007/978-3-319-56808-9_5

Table 5.1 Some categorised uses of bamboo (Gielis 2002)

Use of bamboo as plant	Use of bamboo as material
Agro-forestry Natural stands Plantations Mixed agro-forestry systems	*Chemical industries* Biochemical products Pharmaceutical industry
Ecology Stabilising the soil Uses on marginal land Hedges and screens Minimal land use	*Energy* Charcoal Gasification
Ornamental horticulture	*Local industries* Arts Furniture Kitchen utensils Houses
	Nutritional industries Young shoots for human consumption Fodder *Wood and paper industries* Strand boards Medium-density fibreboard Laminated lumber, paper, rayon, and parquet

roofs for housing, trusses, bridges, and towers. Table 5.1 provides a summary of diversified bamboo utilisation. However, bamboo culm is a vulnerable material and needs to be treated before it is used for any engineering works, and proper treatment can prolong its life span for years. The use of the whole bamboo culm for construction requires appropriate design techniques and some level of skill to bind the round culms together (Janssen 2000; Minke 2012). The handicraft industry has developed well in some countries where bamboo resources are available. The countries in Asia and Africa include China, India, Brasil, and Colombia, as well as Ghana, Kenya, Nigeria, Ethiopia, Japan, while many others also make use of this resource profitably.

From Table 5.1, it can be seen that bamboo can be used in a variety of applications in the engineering and non-engineering fields.

5.2 Engineering Applications

The engineering applications of bamboo, as a base material in the production and manufacturing industries, rely basically on its mechanical properties. Bamboo, as a material, plays a significant role in manufacturing and designing works. It's mechanical and physical properties qualify it as a suitable alternative material for use in the building, construction, automobile, aerospace, art and design, food and beverage, textile and garment, and pulp and paper industries. Bamboo, for centuries past, was referred to as a poor man's material in localities where it is available, especially in Africa, Asia, and Latin and South America. However, in recent years, bamboo has seen a dramatic growth in applications for structures and many other products that are eco-friendly and renewable (Janssen 2000). Bamboo's multi-functional properties prove to be excellent and make it a sustainable natural resource material for engineering works (Naxium 2001).

The engineering applications of bamboo can be categorised into two groups, ranging from domestic to industrial applications. It also finds applications in handicraft and architecture products made manually through processing relatively small volumes of bamboo. The industrial processing of bamboo products involves semi-mechanized and mechanised processing of large quantities of bamboo culms into the required dimensions and applications. A variety of engineered products is manufactured from bamboo, with premium bamboo parts including high-value products such as floor tiles, furniture, toys, plates, cups, bowls, blinds, mats, musical instrument, weapons, chopsticks, and trays. Its applications also include wide use in the field of electronics, i.e. for the production of electronic frames or cases for electronic gadgets. For example, bamboo has been used to produce case or frames for USB pen drives, laptops, radios, televisions, phones, remote controls, mouses, keyboards, and watches. Additionally, it is utilised in the production of paper, charcoal, chipboard, laminate and fibre composite, and so forth. Bamboo products have an excellent advantage in load-carrying components, where its mechanical properties, such as compressive strength, modulus of elasticity, and many others are essential to support structures. Bamboo's application is very common in areas where load applications are needed. Other products derived from bamboo include veneers, panel boards, decorative articles, household appliances, and other building materials. The traditional applications of bamboo for centuries has been, and still are, for walling, fencing, water pipes, fishing rods, musical instruments, roofing, temporary structures, mats, baskets, and others. In most rural farming areas where bamboo is available, its typical applications include temporary housing, temporary interior and exterior decorations, roofing of buildings, antenna poles, scaffolding in construction sites, and as a beam support in construction sites. Some engineering applications of bamboo culms are further described in detail in the following sections.

5.3 Bamboo in Construction

The bamboo culm is a dynamic material and a suitable alternative to wood in the construction industry because it possesses several economic and ecological advantages. The mechanical properties associated with bamboo increase with the thickness of the fibre walls, making it an alternative to wood for the construction industry. The mechanical properties of bamboo culms make them an appropriate choice for load–bearing structures. There are several uses of bamboo in the construction sector such as for bridges, scaffolding, boats, houses, walls, parts of wind-turbine blades, truck bodies, bicycles, furniture, tricycles, and panels (Sharma 2014; Amada and Untao 2001; Lobovikov et al. 2007). In the construction industry, bamboo culms and bamboo-based products are mostly utilised for load-bearing but also non-load-bearing applications. The vulnerable nature of bamboo requires that the bamboo and its products be treated very well with preservatives, for safety reasons and also to enhance its life span. When bamboo culms are utilised indoors, the safety procedures are less significant. However, it is imperative to protect the material with chemicals as a preservative against mould, mildew, fungus growth and destructive insects, at least in the tropical and subtropical countries, when it is used for interior design, furniture, flooring, etc.

5.3.1 Bamboo for Housing

The long tradition of using bamboo for building has spanned diverse cultures and traditions, among people of various geographical areas across the globe where this resource is distributed. While t is an economical material for the construction of houses, temporary shelters, scaffolding, and in many other applications, bamboo usage for construction works has mostly been in rural communities of developing countries. In these societies, most of the buildings are designed and constructed based on the the traditions of the rural dwellers applying their skills and methods. The methods and tools used in these constructions are very simple and straightforward, as they can be utilised by the skilled and the unskilled worker.

Some general uses of bamboo in domestic works include school buildings, market shelters, the construction of restaurants, reinforcement in concrete for buildings, and scaffolding in communities where it is available. Bamboo is an excellent construction material for building houses in disaster-prone areas due to its potential to survive seismic events. As a lightweight and elastic material, it has a high potential to resist earthquake shocks (Tistl and Velásquezgil 2001). Also, tt possesses excellent insulation properties because it is hollow in nature.

The culm is utilised for both small houses and larger buildings, such as garages, sheds, protected yards, canopies, smaller, and larger industrial structures. It is also utilised for making short- and long-span bridges as an original load-bearing material. Bamboo culms are treated very well with chemicals and proper protection against fungi and insects, as well as rain. This treatment is known as post-harvest treatment. The use of bamboo for structural works is not limited to the bamboo culm alone. It also has an extensive range of applications, which involves use as an entirely raw material that is then transformed into engineered products.

Bamboo-engineered products can be an alternative material in the fields where wood-engineered products are being utilised, such as various constructions and building works. The use of engineered bamboo products has recently expanded into the construction industries where structural engineers are utilising it effectively for their works. The fast development of various types of bamboo products in the engineering industry, as well as the manufacturing sector, has now led to additional competition with wood products, especially in areas where wood resources are limited and bamboo is available for the construction of buildings.

5.3.2 Bamboo for Walls

Bamboo has in recent years been adopted thoroughly in the construction and building industry, as well as for walls and partitioning. Bamboo is light in weight, so bamboo walls and partitions can be erected with fewer people, both skilled and unskilled. Bamboo is used for primary elements such as columns and beams, and generally as part of the structural framework. It can carry the intrinsic weight of the building and the loads imposed by the occupants, as well as stresses arising from the weather. An infill between the framing members is required to complete the wall. The purpose of the infill is to protect the culms against rain, wind, and animals, to give privacy, and to provide in-plane bracing to ensure the stability of the general structure when subjected to horizontal forces. Bamboo used for walls can be prepared or handled using various methods such as:

- The whole bamboo culm is halved or stripped.
- The split bamboo culm is woven into columns or mats and fastened together, after which mud is applied on both sides of the column to cover the split bamboos.
- The mud covered panels, depending on the species, are further limed or cement plastered to achieve a smooth surface finishing of appearance and hygiene.

5.3.3 Bamboo for Flooring

Traditionally bamboo culm is used as a flooring material atop compacted earth, with or without a covering of bamboo matting. The procedure employed in the installation of the bamboo floor is to raise the floor material above the ground, allowing for the creation of a stilt type construction (Sharma et al. 2014). The purpose of this procedure is to improve comfort in the rooms as well as provide hygienic flooring to its occupants. However, the exterior part of the earth floor is occasionally made firm with the application of unpolished bamboo boards manufactured by flattening the solid bamboo culms into strips. Bamboos utilisation as a floor material includes the following techniques:

- **Small bamboo-culm flooring**: Culms of bamboo are nailed together as a unit or directly bound together using ropes or wires in their production.
- **Split-bamboo flooring**: This product is produced from the splits of bamboo culms of several lengths of strips which are bound with the addition of adhesives

and then pressed together using an external force after which it is allowed to cure before its use on the floor of rooms.

- **Flattened-bamboo flooring**: This type of floor tile is produced by splitting green bamboo culms, removing the diaphragms from the culms, and rolling and then flattening them to the required product measurement. The split culms are made into thin boards that are then put across each other to form beams, which finally are secured by fastening or tying. It is then partitioned with cement mortar for hygiene and the comfort of the users.
- **Bamboo-mat flooring**: This product involves splitting bamboo culms into thin strips varying in size from 5 to 6 mm or 10 to 15 mm, and with a thickness ranging 0.6–1.2 mm. These slices are then woven into mats of various sizes, and hot-press plates are applied to it as a chemical curing process takes place. After drying the mats to a moisture content of 6–12%, glue or another adhesive bonding material is then applied to ensure enough bonding between the overlapped areas of the mats. The most common adhesives for bonding bamboo mats include phenolic resins.
- **Bamboo plastic flooring**: This type of floor tile uses advanced technology in which bamboo fibre in the raw material state is compounded with plastic as the core material in the manufacturing process. It possesses high water resistance as well as high dimensional stability properties when compared to original floor tiles.
- **Laminated bamboo flooring**: This type of floor tile is produced by splitting the culms of the bamboo into thick strips. These strips are then coated with resins, arranged into units of three or more layers, and then put in a hot press machine to bind the layers together, after which they are allowed to cure completely. The final product is then trimmed to the required shape and varnished. The natural grain of the bamboo structure is clearly visible after varnishing and is very attractive. This type of floor tile has higher resistant to moisture and pressure and has a unique appearance, as compared to soft or hard wood. Bamboo possesses a flexible natural property when used as a flooring material, can last longer than soft- or hardwood, and is a less-expensive material.

5.3.4 Bamboo for Roofing

The use of bamboo in producing sheet roofing is basically for the purpose of protecting structures and occupants from the extremes of weather—including rain, the sun, and the wind—and to provide shelter in a clear and usable space beneath the coverings. The bamboo culm needs to be robust enough to resist the considerable forces generated by wind on roof coverings. Bamboo culm is ideal as a roofing material because it is strong, resilient, and lightweight. Most often bamboo culm used as a structure for roofing is comprised of purlins, rafters, and trusses.

Fig. 5.1 Corrugated roofing sheet. *Source* Guadua Bamboo (2016)

- The simplest method for roofing consists of a bamboo purlin and beams supported on perimeter posts. The bamboo culm is split into two halves and then laid convex side down, edge-to-edge, spanning from the ridge to the eaves. The second layer of halved culm, laid convex side up, is then laid to cover the joints.
- Corrugated-bamboo roofing sheets are made from several layers of woven mats of bamboo culm which can be used for roof coverings. The product, woven into mats, is impregnated with an adhesive resin, dried, and heat pressed together between two specially-designed corrugated pressing plates to form strong, reliable sheets of bamboo, which are lightweight (Fig. 5.1). This grooved sheet of bamboo has excellent insulation properties. The sheets can be produced in a range of sizes to suit particular requirements and can easily be trimmed for special applications. These roofing sheets are durable, stable, and resistant to pest attack, fire, and severe weather.
- Bamboo roofing sheets are an excellent alternative to corrugated metal, plastic, or asbestos roofing sheets. The products manufactured from the culms have applications that cut across a wide range of roofing needs for temporary or permanent structures. Roofing sheets from this product make less noise during rains and also remain cooler during exposure to the sun, when compared to other sheet materials. Some advantages of this type of sheet roofing are that it possesses a pleasing aesthetic appearance, high resistance to fungal and insect attacks, can be easily worked, and has excellent bonding strength.
- Another usage of bamboo culm for roofing sheeting involves a layer of bitumen inserted between two woven mats of bamboo to form a semi-rigid panel. The mats can be formed into beams 200–250 mm from centre to centre, while the surface is finished with a bituminous or rubberized weatherproof coating.
- This form of roofing sheet, finished with cement coated with or without the addition of organic fibres, is traditionally applied to bamboo culms that are used to make a stronger roof covering. Trusses using bamboo culms are made in various diameters, ranging 40–100 mm.

5.3.5 Bamboo for Scaffolding

Bamboo has been adopted as one of the most reliable materials for the construction of scaffolding, due to its weight in relationship to the load-bearing capacity, irrespective of how tall the buildings. In the making of scaffolding, only lashed joints are used. The vertical and horizontal bamboo culms used for scaffolding are almost entirely joined using soft lashing. This method of making scaffolding has a significant advantage because the joints can be re-tensioned to the desired angle without any difficulty and also can be released quickly. In Asia and South America, bamboo scaffolding is even used in the construction of tall buildings.

5.3.6 Bamboo for Interior Work

Bamboo for interior decoration has been used for ages in areas where it is available. Its applications for decoration ranges from split culms, strips, bamboo-culm panels processed in a rectangular form, and woven bamboo mats, etc. Bamboo culm has a unique visual appearance when used for interior decoration and can be distinguished from other soft and hard woods and their products. The use of bamboo culms for interior decoration requires special skills and tools to secure, bind, and mount the parts together before fixing them to the walls or ceilings. It is used in ceilings, roofs, and flooring attracts people's attention and seems to affect their mood. Currently, bamboo and bamboo products are gaining more popularity in the interior-decoration industry across the globe, where they are used for flooring, ceilings, wall cladding, and partitioning, etc.

5.3.7 Bamboo for Furniture

Bamboo used for making furniture consists of diverse shapes. There are two main types of bamboo furniture making, namely: traditional bamboo furniture which uses whole bamboo culms as a primary material and prefabricated bamboo products using advanced laminated bamboo, veneers, and other engineered bamboo panels. The process whereby the entire bamboo culm is used for manufacturing of furniture requires some specific skills and techniques. The bamboo culms require specific cutting and shaping techniques to enable connections of the bamboo furniture parts (Fig. 5.2).

Some of the methods available include: interlacing with split bamboo, natural fibres or ropes, bolting, pinning, or using metal connectors. For ages, traditional bamboo furniture has been produced in areas of its availability by using the natural round or split culm of the bamboo. The furniture products made from whole or split

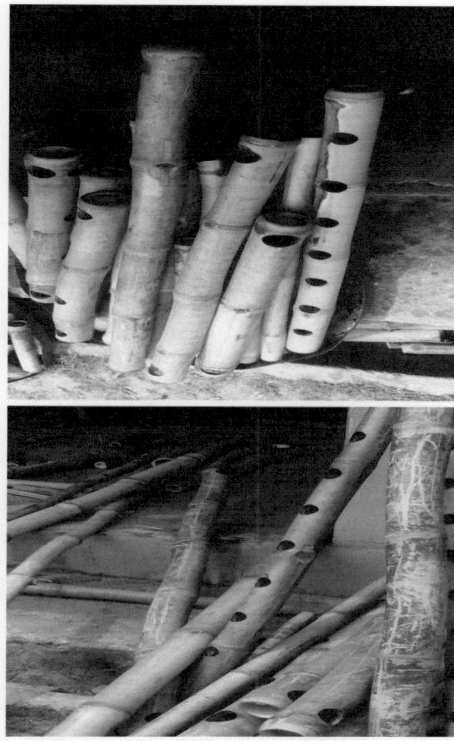

Fig. 5.2 Special cutting and shapes for bamboo furniture

bamboo culm range from tables, doors, windows, book shelves, window and door frames, chairs, beds, stools, and kitchen cabinets, among many others (Fig. 5.3). However, for greater durability and longer life span, they are treated with chemicals (Dransfield and Wiidjaja 1995; Rubio-Luna 2007).

The application of glue onto bamboo furniture joints is very limited due to the curvilinear nature of the bamboo, and also because the skin does not allow proper bonding, even when high-tech gluing systems are applied. When furniture is made from bamboo culms having their skins intact, painting or spraying is not needed because it has a hard, smooth, and impermeable surface providing sufficient protection against dirt and wear. Nevertheless, there is a need for protection against insects and fungi that attack bamboo and its products. Bamboo has a high sugar and starch content and, due to this, mould easily develops on its products such as furniture; this is especially the case on cross sections and inner surfaces when high relative humidity and elevated temperatures are predominant.

Furniture manufactured from engineered bamboo products have properties and features similar to those of wood-based furniture. Bamboo furniture of traditional design may be transported in small flat packs to the spot of assembly that does not require any sophisticated tools. Like original whole-bamboo furniture, engineered

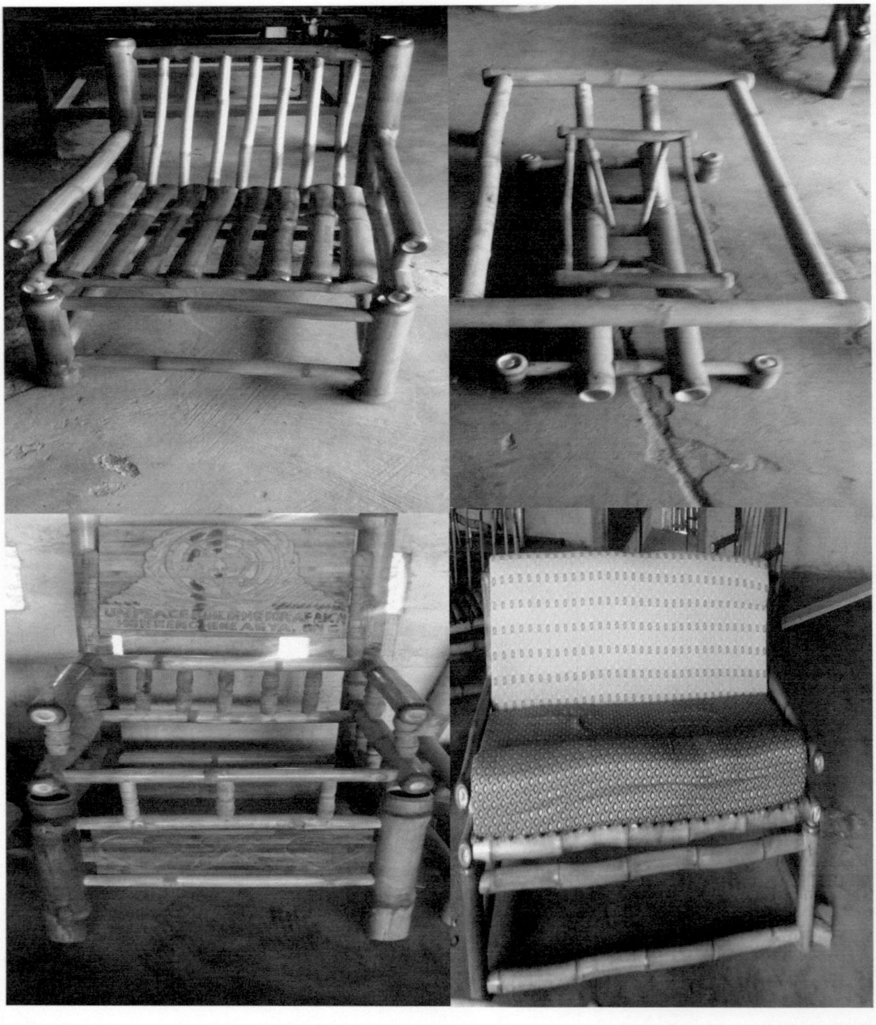

Fig. 5.3 Examples of traditional bamboo furniture

bamboo furniture needs to be preserved and surface treated to prevent damage from fungi, insects, and daily use (Fig. 5.4). Various kinds of oil, waxes, lacquers, and paints are used in preserving the surface. The gluing and connecting techniques are similar or equal to those used for wood-based furniture, with preference given to certain types of glue. Water or weather-resistant glues are preferred to avoid failure of glue lines. New designs have overcome many of the problems of traditional bamboo furniture, including labour-intensive stresses, varying drying quality, and vulnerability to insects and fungi. Advances in the treatment of culms helps to retain the distinct physical, mechanical, chemical, and aesthetic features of bamboo, while

Fig. 5.4 Engineered ply bamboo

current trends of processing and manufacturing of furniture no longer depend on artisans but rather on industrial-scale production lines at high-speed and high-quality levels.

5.3.8 *Industrial Bamboo Board*

Through the industrial processing of bamboo, almost everything that can be made from hard or soft wood can also be manufactured with industrial bamboo materials. The industrial processing of bamboo, in particular, the lamination of bamboo strips into boards (Plybamboo), is mostly applied in flooring, furniture board, and veneers. Moreover, board materials can also be produced from woven bamboo mats. Plybamboo is another type of product produced from bamboo industrially and is available in various colours and sizes. Over the past few years, many innovations in the field of production technology have led to the development of new industrial bamboo materials with various different properties and potentials, such as bamboo mat board (BMB), bamboo particle board (BPB), and many others, to briefly cite the various experiments into bamboo composites. BMB is made from thin bamboo strips or slivers woven into mats to which resin is then applied and pressed together under high pressure and high temperature, which makes the mats into extremely hard boards, which can be put into moulds and processed into corrugated boards.

- **Strand Woven Bamboo (SWB)**: A relatively new bamboo material is made from thin and rough bamboo strips that are glued under high pressure in moulds and formed into beams. Strand bamboo is used mostly for indoor and outdoor decorations, with a high hardness (2800 lbf) and density (1080 kg/m^3) due to the compressed bamboo strips used in combination with a high resin content. An interesting feature of SWB is that there are no high requirements for input strips, which means that, unlike the production of Plybamboo, a large part of the resource can be used, thereby utilising the high biomass production of bamboo to the maximum. Due to the compression and the addition of resin, SWB has a very high density (approximately 1080 kg/m^3) and hardness that makes it a material suitable for use in demanding applications (e.g. staircases in department stores). Recently, a new higher resin-content version of SWB has been developed for outside use, which could make SWB a suitable alternative for rare tropical hardwood species.

- **Bamboo Particle Board and Bamboo Plastic Composites**: These types of board are still in the early stages of development. Their processing techniques are based on existing techniques from the wood industry and are not yet widely available commercially.
- **Bamboo Mat Board**: This type of board is made from thin bamboo slivers and strips that are woven into large mats, which can serve as input for the production of various boards, including bamboo mat board (BMB) which can be pressed into moulds of various shapes (including corrugated boards).
- **Chipboard**: Bamboo chipboard is produced from dry bamboo shavings, mixed with a binding element that provides water resistance to the surface using a waterproof agent on the board and hot-pressed at a given temperature with the proper pressure. The culms are taken from green forests, as well as farm plantations, to produce the product, and its production is an effective way to raise the utilisation ratio of bamboo resources. To produce bamboo chipboard, a water-soluble phenol resin is used. Chipboard has a higher modulus of rupture, higher modulus of elasticity, higher water tolerance, and has a lower moisture expansion in thickness, as compared with wood chipboard. Chipboard being an engineering construction material is applied to conventional concrete works. To improve the board's application, lesser-known species of culms and other residues of the bamboo are processed to produce the chipboard. The manufacturing process involved in producing chipboard is the about the same as the design and technology for producing wood particle-board through rolling, cutting, chipping, re-drying, gluing, spreading, and hot-pressing. The utilisation ratio of raw material for chipboard production is high, for example, 1.3 tonnes of the raw material 1 m^3 of chipboard can be produced (Shi and Walker 2006). The manufacturing process of bamboo chipboard employs the same technology as wood particleboard. The usage of phenol-formaldehyde resin for the manufacturing of the product is comparatively advantageous in strength since MOE has a low expansion rate of water absorption.

5.3.9 Bamboo for Handicrafts

The special features associated with bamboo-culm handicrafts made from this raw material can easily be recognised. The round shape, the smooth skin, the characteristic colour, and the texture originates from the vascular bundles embedded in parenchyma tissue, and the node or internode sequence are typical features of bamboo handicrafts. Bamboo handicrafts comprise all kinds of useful items, such as kitchen tools and accessories to assist cooking and the presentation of food, bowls, mugs, bird homes, among many others.

5.3.10 Bamboo for Musical Instruments

Traditionally, natural bamboo culm has been used for the manufacturing of musical instruments for centuries. Bamboo culm has many applications regarding its usage for instruments of all classes like wind, percussion, cord instruments, and the cords itself can be made exclusively from this one material. The hollow structure of a bamboo culm provides a characteristic sound relating to the music of many ethnic groups. Bamboo flutes have existed from time immemorial. The bottom part of smaller culms requires only a few finger holes and a mouthpiece to produce an excellent flute. Lengths of the culms are used as resonators below the bronze keys of xylophone-like instruments.

5.4 Engineering Bamboo Products

In recent years, various engineered bamboo products have been developed and are available in the market for engineering purposes. The term "engineered bamboo products" covers a broad range of bamboo derivative products. Common to all engineered bamboo products is that the culm is broken down into smaller units such as lamellae, splits, strands, and particles which are further processed by binding or fixing the units together with adhesives, plastics, or other methods of fixation to form a composite material with uniform properties. Engineered bamboo products are used in a variety of applications, including handicrafts, furniture, interior design, flooring, and structural and automotive applications.

5.4.1 Laminated Bamboo Lumber

In producing laminated bamboo lumber, the round-shaped hollow bamboo culm needs to be converted into rectangular elements. In this process, first, the bamboo culms are split into pieces after which they are pre-planed to remove the skin and the inner layer. The splits are planed on four sides to obtain uniform rectangular lamellae, which can be further processed into laminated bamboo lumber. Resins are applied to the lamellae, after which the lamellae are assembled and placed in presses. Laminate bamboo lumber with an extensive range of dimensions can be produced by glueing together smaller pieces which have been manufactured to the required measurements. Laminate bamboo lumber can be used to produce window frames, as a structural building material, and joinery. It is always appropriate to protect the bamboo laminate lumber and its products against moisture, fungi, and insects before being used, depending upon the species.

5.4.2 *Bamboo Veneer*

Bamboo veneer can be produced on a peeler lathe as culm sections are rotated against a knife and nose-bar assembly. The process of peeling of bamboo culm by this method is difficult due to the small diameter and the hollow tube-like structure of the equipment. The most common technique used in peeling is the slicing technique. In this case, the planks of laminate bamboo lumber are sliced on horizontal or vertical lathes perpendicular to the grain direction or longitudinal lathes along the grain direction, whenever bamboo slices are glued together to form larger pieces for veneer manufacturing. Assembly of a bamboo slice of various colours produces a broad range of decorative veneer production. The thicknesses of bamboo veneers range between 0.15 (micro-veneer) and 1.5 mm. Bamboo veneers are very fragile because of their low tensile strength perpendicular to the grain direction. The tensile strength property perpendicular to the grain can improve fragile nonwoven cloth-type support layers that are attached to the backside. Bamboo veneer is a product with high added value that is used as a surface layer in coating processes for other bamboo or wood-based panels. The coating process improves the visual appearance, by presenting a bamboo composite look.

5.4.3 *PlyBamboo*

Plybamboo is a one category in the broad range of bamboo-based panels, which is produced from layers of bamboo veneers with certain desired thicknesses. The veneer sheets are assembled so that layers have their grain direction rotated 90° about adjacent layers. The thickness of veneers can vary over a wide range. Large layers can be produced by glueing smaller sheets edgewise together. Because bamboo culms have high rigidity, they can hardly be deformed to fill up the blank space between strips even under high pressure. This results in the lowering the modulus of rupture (MOR) and its adhering strength.

Plybamboo can be characterised as high quality with low weight, high rigidity, high friction coefficient and does not rust. In addition, the manufacturing process for Plybamboo was found to be less laborious and to consume fewer adhesives than other types of composites. The strength, wearability, and rigidity of Plybamboo are higher than those of ordinary plywood. Thus, Plybamboo has a broad prospective in the automotive and building industries and also for engineering construction (Mahdavi et al. 2012). To provide the raw material for Plybamboo, not only veneers but also splits, lamella, and slivers are utilised for the formation of middle layers.

Depending on the inner structure and alignment of the inner layers, a broad variety of different Plybamboo products can be produced. In producing the Plybamboo, the middle and surface layers of the bamboo culm are usually split into sections and then are roughly planed to remove the skin. The splits are then boiled, bleached, and anti-mould or anti-pest treated. The heat treatment can adjust colour (often, but not correctly, termed carbonising) during the drying process. The dry splits are then cut to equal width and thickness, sorted, aligned, and assembled to form layers, which are glued together inside presses or plate hot presses. After thickness calibration, several sheets or plies (with identical or different structures) are assembled crosswise similar to plywood forming panel-type Plybamboo products with low anisotropy.

Another alternative for producing the middle layers of Plybamboo is to use curtains or mats. Curtains are formed by connecting splits using thin filaments. In this case, the edges do not have to be aligned. For producing bamboo mats, splits have to be cut into thin slivers (0.8–1.5 mm). For high-quality Plybamboo panels, the thickness variation of the slivers should be as low as possible, certainly not more than 1 mm. Slivers have to be woven into mats which are cut to predetermined dimensions before being used to form middle layers in Plybamboo panels. Many different forms of Plybamboo are on the market now. They can be applied as panels for heavy-duty containers, flooring materials, and decorative materials for furniture production or interior design.

5.4.4 Oriented-Strand Board

To manufacture bamboo into oriented-strand board, the culm is split into thin strands and then cut into sections which are introduced into a flake. Ideally, the cutting direction must be along the radial axis so that the outer dermis is always located on the thin side of the strands. Dermis located on the wide side of the strand cannot be aligned properly and will result in poor internal bonds in oriented-strand board. Strands are dried in rotary drum dryers and aligned in mixing units. After alignment, the strands are scattered onto a moving belt. Normally, strands are aligned by the scattering unit and by rotating the next scattering unit by 90°. With several adjacent layers, plywood-like panel properties can be achieved. A normal mat-forming station comprises three or five scattering units. The strand mat is either transported to a continuous hot press or, after separating into sections, to an 8–15 h daylight hot press (Malanit 2009).

5.4.5 Bamboo Particleboard

Recently, the production of bamboo particleboard in the wood industry is on the increase, as bamboo is inexpensive and its maturity is shorter than wood. In the manufacture of this board, particular attention has to be given to the fact that bamboo, depending on age and time of harvest, contains considerable amounts of sugar and starch. If the raw bamboo is not treated well, it may be negatively affected by fungi and insects during storage. Also, if the finished bamboo products are produced under humid conditions, fungal attack can endanger them. Therefore, precautions need to be taken in preserving the bamboo and its particleboard from fungi and insects.

The bamboo culms, when cut into smaller pieces, are fed into a knife ring. At this stage, the bamboo culm is transformed, preferably, into longish particles. When this has been done, then the particles are separated in a sieving station into coarse-and-middle-layer particles and fine surface-layer particles. The particles are then dried to low moisture content, and the particles are vibrated in a blending unit. The moisture content of the surface layer particles is adjusted to approximately 10–14%, whereas the moisture content of the core layer particles is kept at 6–8%. The particle mat is formed by scattering first the lower face layer, followed by the coarse core layer and an upper-face layer. The mat is then transported to a continuous hot press where the mat is compressed and heated until the resin is cured. During hot pressing, the typical U-shaped density profile is formed, with high density in the face layers and lower density in the core layer. During hot pressing, part of the moisture contained in the face-layer particles is converted into vapour which travels to the core where it assists the heating process. Moisture distribution across the panel thickness should be more or less uniform at the end of the pressure cycle.

The endless particleboard belt is cut into sections when it leaves the press. Sections are introduced into a cooling star to allow the panels to cool down to a temperature close to ambient temperature. Finally, panels are calibrated in a sanding process. Panels are then stacked to form piles for transport to the customer or further cut into smaller units according to the client's requirement.

Due to its properties, particleboard is typically used for the manufacturing of furniture. Its high-density face layer allows veneering and coating with melamine impregnated paper. Edges can be edge coated so that the particleboard core is not visible in the final product. The visual appearance of bamboo particleboard is identical to wood particleboard.

5.4.6 Nonwoven Strand-Bamboo Products

Nonwoven strand bamboo is a new type of bamboo product which requires a more advanced manufacturing process than does nonwoven (compressed) strand bamboo. This new product is found on the market as a plank, scantling (either four-sided planed, profiled), tongue and groove element, parquet element, or in other forms.

The bamboo culm is crushed by splitting the bamboo culm into segments that are fed through motor-driven rollers or a crusher, which crushes the bamboo into a type of mat. In some cases, the skin is removed using a preplaning process. In the longitudinal direction, vascular bundles and surrounding parenchyma tissue are separated without destroying the vascular bundles. Bamboo mats are boiled to reduce sugar and starch content, dried, and subsequently resinated by dipping the mat into a resin bath (usually phenol-formaldehyde). The soaked mats are then transferred into pressing moulds. In a hot press, the bamboo mats are compressed and densified to approx. 1.0 g/cm^3 or even higher while the resin is cured. After cooling and conditioning, the bamboo elements are planed, cut, or profiled to the desired dimensions. Nonwoven strand bamboo or compressed bamboo products exhibit extreme hardness, very high flexibility, and compression strength, and are comparable with or even better than the strongest wood species.

Due to its high density and high resin content (15–20%), moisture uptake of nonwoven strand bamboo is slow. Nevertheless, long-term exposure to humid climate or liquid water will lead to moisture absorption and swelling. During swelling, the compressed bamboo will break the resin bonds. As a consequence, the swelling of compressed lignocellulose material is not fully recoverable during shrinkage, which leads to unwanted deformations. For this reason, nonwoven strand-bamboo products must be thoroughly protected against moisture uptake, either by coating with water-repellent surface treatment or, even better, by not exposing them to a long-term humid climate.

5.4.7 Bamboo Plastic Composite

Bamboo plastic composite (BPC) is a product derived from the culm of a bamboo plant. The bamboo culm is ground into flour-like powder or very fine particles, which are mixed with plastic materials (normally PP or PVC) in a compounder. Here, up to 60% (weight basis) of woody particles are thoroughly mixed with 40% or more of plastic matrix material. The resulting compounded material is used in extrusion or injection processes. When bamboo culm is ground, it can easily be used instead of wood. Typical BPC products are extruded planks for decking, which are highly prized in the market for construction works. Due to its water-repellent properties, BPC planks are substitutes for pressure-treated wooden planks or wooden planks produced from wood species with high natural durability.

5.4.8 Bamboo Chipboard or Flakeboard

The bamboo chipboard is produced from bamboo shavings as elementary units, which are dried, mixed with a certain amount of adhesive and waterproofing agent, spread, shaped and hot-pressed at the required temperature and with the required pressure. Flakeboard is made up of small-sized bamboo waste or culms. The raw material used in manufacturing bamboo chipboard comprises all the parts of the bamboo. The production of chipboard using bamboo is an effective way of achieving complete exploitation of the plant. The chipboard made of bamboo is produced using water-soluble phenol resin because it has higher water tolerance, higher modulus of rupture, modulus of elasticity, and lower moisture expansion in thickness as compared with wood chipboard. The manufacturing process and the related designs follow the technology of wood particleboard. They are accomplished through rolling, cutting, chipping, re-drying, gluing, spreading, and hot-pressing. They are used as a material for engineering construction in the building industry, and also find applications for making ordinary concrete forms.

Usually, the bamboo culms of small diameter and lesser-known species are those processed with other parts to manufacture bamboo chipboards. All leftovers of bamboo culms and residue of bamboo cuttings are treated together to manufacture bamboo chipboards (Shi and Walker 2006). The utilisation ratio of raw material for chipboard production is high, e.g. from 1.3 tonne of raw bamboo material, 1 m^3 of chipboard can be produced. This product can be strengthened by adding a bamboo curtain or bamboo mat to the surface. The main drawback with flakeboard or chipboard is the difficulty in bonding and so is the major problem when it is shaved. However, the adhering quality of bamboo chipboard is high.

5.4.9 Elastomer-Based Biocomposite

The use of bamboo fibres has expanded into use in elastomer composites, as a newly viable, alternative filler reinforcement. The short bamboo fibres are employed in a matrix compound due to the considerable processing advantages, improvement in certain mechanical properties, and economic considerations. The addition of short bamboo fibres to elastomer-polymer matrix, especially natural rubber (polyisoprene), produces great mechanical performance of the manufactured composites (Ismail et al. 2002; Visakh et al. 2012). It has been found that the bonding agents (silane, phenol formaldehyde, and hexamethylenetetramine) play a major role in obtaining good adhesion between bamboo fibres and the matrix. The resulting material exhibits composite properties, for instance, fatigue strength, modulus, elongation at failure, creep resistance over the particulate filled matrix, hardness, and cut, tear, and enhanced puncture resistance. Applications of elastomer composites include tires, gloves, V-belts, hoses, and complexly shaped mechanical goods. As for tire manufacture, carbon black has been extensively used (Muller and Rebelo 2015).

5.4.10 Thermoplastic-Based Bamboo Composite

Bamboo fibre reinforcement includes thermoplastic polypropylene matrices which are broadly applied to produce high-quality products for engineering application

(Bansal and Zoolagud 2002). Bamboo strips have higher cohesive strength than the extracted bamboo fibres from various types of bamboo. For this reason, bamboo strips are reinforced with nonwoven polypropylene aiming to produce ultra-light-weight unconsolidated composites (Siriwan and Suthamnoi 2009). The nonwoven web enables these materials to be reinforced in their native form (Abdul Khalil et al. 2012; Khanam et al. 2011) and to utilise the unique properties of the reinforcing materials. It has been found that bamboo strips-polypropylene (BSPP) composites have improved properties, including high flexibility, high acoustical properties, and good sound-dampening attributes that makes it suitable and an ideal raw material to replace fibreglass. BSPP has application in the manufacture of automotive headliner substrates. Several components can be manufactured using biocomposites, such as door inserts, parcel shelves, and load floors for automotive vehicles, as well as walls and profiling for buildings, and other engineering applications. A study on the effect of bamboo-charcoal addition to the thermoplastic polyolefin polymer reveals that bamboo charcoal has many pores in its structure that make it an excellent medium for preventing static electricity build up and absorbing volatile chemicals (Prasad and Mohana 2011). Bamboo charcoal has been chosen as a promising material to enhance the water absorption and electrical conductivity of polyolefin.

5.4.11 Thermoset-Based Bamboo

The potential of, and interest in, bamboo used in thermoset composites follow the same trend as thermoplastic composites. The effects on bamboo fibre-reinforced polyester matrix have been evaluated through various tests, e.g. tensile and flexural properties (Mohan and Prasad 2010), dielectric properties (Wong et al. 2010), and fracture properties (Hongyan et al. 2009). In addition, the influence of moisture absorption during storage and manufacture of bamboo-fibre-reinforced vinyl-ester composites has also been analysed (Nirmal and Jamilhashim 2012). In the same manner, bamboo-fibre-reinforced epoxy composites were subjected to wear and a frictional environment to ensure widespread acceptance in many applications (Tong et al. 1998). It was noticed that wear resistance was superior when the fibres were orientated nonparallel to the sliding interface (Corradi et al. 2009). In other tests, bamboo-strip epoxy composites were found to be promising materials to be applied in the global marine sector (Ismail et al. 2001), and bamboo boat hulls have been produced using a vacuum-bagging and compression-moulding process. There have been several products manufactured and tested, and these products were found to exhibit excellent mechanical properties, including good material endurance and resistance to the marine environment. Exploitation of bamboo-epoxy composites was further extended to the manufacturing of surfboards.

5.5 Paper and Textiles

5.5.1 Paper

Bamboo usage for paper production has been around for centuries in some Asian countries, including China and India. In China, handmade paper from bamboo began more 2000 years ago. The inner parts of the bamboo were beaten into a pulp and used in paper production. Today, the annual output of bamboo in India is about 3.23 million tonnes, and the paper industry consumes more than half of this production. Pulp mills using bamboo as the raw material exist in most Asian countries. In India, 30–35 factories make paper from bamboo pulp. The present world production of pulp from bamboo is approximately 1.5 million air-dried tonnes, corresponding to an annual consumption of 7–8 million tonnes of raw bamboo.

The fibre length of bamboo varies from species to species and with the intermodal length. Mean fibre-length values for species range from 1.5 to 4 mm, fibre diameter range from 11 to 19 μm, lumen diameter from 2 to 4 μm, and wall thickness from 4 to 6 μm. The fibres of bamboo are well suited as a raw material for pulp production. They are thinner than wood fibres, which contributes to the desired smoothness and flexibility. The high cellulose content implies suitability for dissolving pulp and for rayon production. Bamboo cellulose contains more impurities than wood. Hence processing is costlier, and the pulp yield in the craft process is similar to hardwood processing.

Since pulp mills require a significant amount of bamboo and its extraction in a tropical climate, it is often limited to 6–8 months per year, and a stock of bamboo for 6–9 months production has to be maintained. A mill producing 100 t per day requires about 43,000 t of bamboo in the yard. During storage, decay by fungi and beetle attack can cause serious damage. In the case of the common brown rot, the yield of pulp is considerably reduced, and the kappa number is so high that the pulp becomes tough to bleach. This is due to a selective attack of the fungi on carbohydrates. Therefore, bamboo, when subjected to brown rot fungi, is unsuitable for pulping. In the case of white rot, bamboo can be used for pulping, but lower yields of pulp and lower physical strength result with more need for bleaching chemicals. During storage for up to 12 months, about 20–25% of the bamboo can be destroyed due to attacks by wood-destroying organisms. Preservative treatment can reduce the storage loss, but sufficient drainage, aeration, and limitation of the size of the stacks also provide protection. Stored culms should have no soil contact, and precautions against fire are necessary.

Because of its anatomical structure, bamboo is a suitable material for pulping. The cooking chemical can penetrate the large vessels and diffuse into the surrounding tissue. Formerly, the culms were first crushed and then chipped. New mills are using chippers only. Due to the hard and slippery skin of the culm, chipping of bamboo requires special techniques. The chips should be 18–20 mm long. In earlier years, pulp mills used a two-stage process for bamboo pulping. First, a weak alkaline solution removed the low polymer carbohydrates, followed by cooking with caustic soda and sodium sulphide. Currently, pulping is carried out mostly in single-stage cooking. Craft process is best suited for producing pulp chemically.

In modern mills with modern cooking technology, bamboo is pulped with 18–20% effective alkali, as NaOH at a low temperature of 142–144 °C produce pulps (50–55% yield) that is well suited for writing and printing paper. Based on kinetic data, regression equations have been developed for controlling the process of craft pulping bamboo for producing pulps with up to a 70% yield. In the production of one tonne of unbleached pulp, two tonnes of clean, chopped bamboo is needed, and the production of one tonne of bleached pulp requires about 2.5 tonnes of air-dried bamboo (4 t fresh). Although nodal portions result in poorer quality of craft pulp, it is not considered feasible to separate nodes and internodes before digestion, since the reduction in yield and quality due to nodal portions seems to be insignificant. Bamboo black liquor obtained by craft cooking contains about 3–4% silica, creating problems in the recovery cycle and requiring partial damping of the lime mud.

The bamboo pulp is more resistant to bleaching than wood pulp because the chemicals cannot penetrate through pits into the cell lumen, as it does in soft wood tracheids. The use of bleach as a lignin-removal agent is less effective on bamboos than on softwoods. Formerly, bleaching was done with a CEHH sequence applying relatively high chlorine charges. In this bleaching sequence, bamboo pulp could not be bleached beyond 78% brightness without severely reducing its strength.

The quality of bamboo pulp is considered to be fairly good. It has a high tearing strength that compares well with softwood craft pulp in this respect. Tensile and bursting strength are lower than those of softwood pulp but still on the same level as those of hardwood-craft pulps. For many purposes, the bamboo pulp is mixed with pulp from other grasses, bagasse, wood, rags, or waste paper. By mixing it with ground wood pulp (60 and 40% bamboo chemical pulp), it can be used in the production of newsprint. The bamboo pulp is suitable for a large variety of paper, e.g. for writing, printing, wrapping, tissues, and so forth. The quality of paper made from bamboo compares favourably with conventional paper made from wood pulp. Bamboo is also used in rayon and cellophane production.

5.5.2 Bamboo Textiles

Bamboo fibres are used in manufacturing the cloth, yarns, and clothing that are components of the bamboo textile industry. The manufactured bamboo-fibre textiles are used in bedding, underwear, t-shirts, socks, blankets, towels, and many other textile products in which bamboo is processed together with other fibres like cotton and polyester to develop a new product of greater and longer lasting value. Bamboo fibre or yarn can be produced through the spinning process to yield various different fibre textures. Moreover, its products prove to have softer qualities as compared to silk and cotton, which are also soft, but much less so than bamboo. Other products manufactured with bamboo fibres include baby diapers and rayon.

The interior of these handmade organic materials are velvet-textured, waterproof, and breathable, producing an always-dry effect on a baby's skin.

Clothing made with these fibres are comfortable to wear, soft, and can be washed like any other cotton shirt. Historically, bamboo has been used mainly for structural elements, such as bustles and the ribs of corsets, but in recent years a range of technologies have been developed enabling the use of bamboo fibre in the textile and fashion industries. Bamboo clothing is made either from 100% bamboo yarn or a blend of bamboo and cotton yarn. Textiles made from bamboo fibres address the goal to develop renewable and sustainable materials. Bamboo is a renewable resource for clothes and other textile applications (Lichtenstadt 1864). The advantages of using bamboo as a raw material for textiles include its renewability, biodegradability, efficient space consumption, low water consumption, organic status, and carbon-sequestering abilities, as well as its sustainability. Bamboo's growth rate is the fastest on the planet, growing as fast as one meter or more per day (Fu 2001; Durst 2006). Bamboo for textile manufacturing provides quicker harvest readiness compared with the 20–60 years for soft and hard woods or trees that are used to make lyocell, rayon, and other regenerated fibre (Rodie 2007).

The manufacturing methods currently being employed in the creation of bamboo textiles include chemically-based (hydro-alkalization and multi-phase bleaching) and mechanically-based (stripping, boiling, and enzyme use) processes.

5.6 Bamboo Bicycles

Bamboo, which is a sustainable, regenerative material, is seen as a sustainable transport alternative. The American designer, Craig Calfee, led *Bamboosero* the first bicycle project for new products developed with bamboo. The Bamboo Bike Project was initiated by Craig Calfee in 1995 and has grown into assisting entrepreneurs in the developing world to make their bamboo bicycles from locally produced bamboos. Developing countries such as Ghana in West Africa are making significant strides in the production of bamboo bicycles as a cheaper source of transportation (Fig. 5.5). These bicycles are put to uses as labour-saving products and as a means of transportation. The bamboo bicycle has become popular because it costs less to manufacture as compared to imported metal bikes, as their manufacture does not require electricity or a significant investment in equipment. The production of bamboo bicycles in Ghana, in collaboration with the local artisans, aims to open markets for export worldwide. Bamboo, due to its high shock resistance, qualifies it as a suitable material for the manufacturing of bicycles, tricycles, and others vehicles for our roads. Producing bamboo bicycles in communities where bamboo is not available is an initiative initiated to help the farmers, artisans, and the consumers, as well as to motivate companies to develop bamboo products using this technology. The role of farmers and artisans in this process is not limited only to crop management, harvesting, and transport, but also includes influence on the final manufacturing of bamboo products. The dominant species used for bicycle

Fig. 5.5 Bamboo Bicycle. *Source* Ghana Bamboo Initiatives (2016)

construction in Ghana are *Bambusa vulgaris*, *Bambusa vulgaris vitta*, among others. The mature culms with long internodes and thicker walls are cleaned, prepared to size, and assembled in the partner units. Currently, there are quite some models of bamboo bicycles on the Ghanaian market.

5.7 Non-engineering Applications

The non-engineering applications of bamboo also span a broad range, e.g. medicine, biomass (charcoal), as a fuel source, as wind break checks, for soil stabilisation, removal of atmospheric carbons, wastewater treatment, and reduction of nitrate contamination. Through the application of science and technology, new methods for the non-engineering use of bamboo have been discovered, and processing bamboo plants into other products has been on the increase in recent years.

5.7.1 Bamboo Charcoal

Bamboo culms find their various application in sectors that include the biomass and biofuel production, such as charcoal and many others. The solid residue remaining in the pyrolysis chamber is the bamboo charcoal, which has a porous microstructure and is primarily comprised of carbon (Ip et al. 2008). Bamboo-charcoal manufacturing can replace the traditional wood charcoal because of its rapid growth characteristics which help to decrease deforestation. Bamboo charcoal is an important source of energy for cooking and heating in many tropical and subtropical regions. The culms by themselves, however, do not form a suitable combustible material because they do not store well, burn fast, and tend to produce a dense smoke while burning if not seasoned properly before usage. Bamboo charcoal offers an alternative for energy storage (Liese and Silbermann 2010). For many centuries, bamboo charcoal has been produced and utilised in China, especially, and also for export to Europe and America in its raw form or processed for other applications (Liese and Silbermann 2010). Besides, bamboo charcoals or briquettes are used for fuel applications. The charcoal is also used in the manufacture of activated carbons (Ip et al. 2008), which is used in clinical toxicology. The charcoal can be utilised as a gas adsorbent and in water and air filters. It also serves as a purifier of drinking water, by removing any heavy metal ions of mercury, nitrate nitrogen, and smells (Lalhruaitluanga et al. 2010; Wang et al. 2010; Tan et al. 2011). Some international organisations are promoting the production and use of bamboo charcoal and its by-products around the globe. They have encouraged a community in Ethiopia to go into its cultivation on about 800,000 hectors of treeless grassland of the lowland bamboo Oxytenanthera abyssinica which has become the only combustible material available for cooking in that community. Bamboo culms can be compared to solid wood as a short-lived combustible material: the pieces burn rapidly and do not maintain heat, so new material (culms) must frequently be added.

In another scenario in Asia, bamboo is referred to as “black gold.” Bamboo charcoal is available as culm segments, as chunks, compressed into briquettes, or in granular or powder form. In its basic forms, it is used for cooking, heating, or smoke-free grilling; however, for the most part, it is further processed into numerous products. International programs such as INBAR (International Network for Bamboo and Rattan) and ITTO (International Tropical Timber Organization) are currently active in promoting the production and use of bamboo charcoal and its by-products effectively. INBAR and the European Union are supporting the use of bamboo charcoal as an energy source instead of burning wood in Ethiopia and Ghana. New industrial capacity for environmentally friendly products will arise through these international programs, especially in rural areas (Fu 2003).

Bamboo charcoal can be used to treat drinking water sterilised with chlorine to remove its residual chlorine and chlorides. These uses of charcoal, especially bamboo charcoal, are attributed to its porous microstructure and large surface area. Bamboo charcoal can act both as a humidifier and as a dehumidifier, depending on the relative humidity, because it releases or absorbs moisture from the environment, hence recommending it for daily domestic household applications (Jiang 2004; Guan 2004). The charcoal made from bamboo is used to manufacture carbon-based composites, nanorods, functional fabrics, silicon carbide, metal reduction and recycling.

5.7.2 Industrial Processing of Bamboo

With the advancement of the traditional methods for industrial processing, whereby the oven contains a conveyor belt or is made with a rotating body, a large quantity of bamboo charcoal, vinegar, and gas can be produced with continuous use of an extrusion press. The culms are broken into chips to fracture the thick outer epidermis. The manufacturing process for bamboo charcoal involves grinding the charcoal pieces, adding water and adhesive, compressing the material into six-sided tubes, and drying. Compressed bamboo charcoal is very hard with fine pores, promoting sustained burning at high heat. The hexagonal shape is advantageous for shipping in cartons since the honeycomb structure is highly stable, sustaining little breakage and taking less packing space. Because of the higher acquisition and maintenance costs, such ovens are used only for the mass production of bamboo charcoal, whose raw material is harvested regionally and supplemented by industrial scraps.

5.7.3 Quality Control

There are no particular international standards for the quality of bamboo charcoal. However, raw density, hardness, electrical resistance, and other properties make bamboo charcoal comparable to many types of wooden charcoal. The "Hardgrove-Index" serves as a measure of hardness by grinding: the lower the index, the harder the charcoal. Industrial analysis examines various characteristics, such as the content of moisture, ash, volatile components, and fixed carbon.

5.7.4 Properties and Utilisation of Bamboo Charcoal

The highly absorptive quality of bamboo charcoal is utilised in various fields. Bamboo charcoal absorbs harmful substances from room air such as formaldehyde, ammonia, and benzene. Its high porosity binds moisture from the humid air and releases it with decreasing humidity. Areas, where bamboo charcoal is applied, include humid spaces like bathrooms and bedrooms, as well as for pillows and bedding. Bamboo charcoal is also mixed into walls, floors, and placed under houses. Granular or pulverised charcoal is incorporated into paper bags, pillows, or mattresses and is woven into outerwear. Depending upon the intensity of its use, the effectiveness is about three weeks; after washing and drying, the material can again be put back into use. Bamboo charcoal binds with dangerous substances such as carbon monoxide, carbon dioxide, benzopyrene, and, above all, nicotine and tar. In China, cigarette filters contain bamboo charcoal which is said to absorb 95% of the toxic substances.

In the kitchen, smoke filters often contain bamboo charcoal. Fruits and vegetables produce ethylene in the refrigerator; charcoal binds these, lengthening the time the products stay fresh. The same applies to odours from fish and meat products. Bamboo charcoal is used to control body odour as well, in the form of shoe inserts and also as an additive in soaps.

- **Electromagnetic (EM) wave protection**: Bamboo charcoal or activated carbon, when placed near electrical appliances such as television, mobile phone, laptop and desktop computers, helps in dissipation of the EM waves.
- **Odour and humidity protection**: Bamboo charcoal or activated-carbon pack product is very effective in removing odours from rooms, cars, washrooms, and kitchens, as well as absorbing moisture, and preventing mildew in shoes and clothing.
- **Skin care**: Bamboo charcoal or activated carbon and vinegar make high-quality ingredients for skin care products, as the former has excellent cleansing ability, while the latter is an excellent skin nourisher.
- **Bath and shower**: Bamboo charcoal or activated carbon added to bath water will give a revitalising hot-spring experience. One can use a bar of bamboo charcoal or activated carbon or vinegar soap to shower. A few drops of bamboo vinegar in the water are beneficial to the body.
- **Refrigerator**: Bamboo charcoal or activated carbon can be used in the fridge to absorb smells and to help to maintain the freshness of food by absorbing the ethylene gas produced by vegetables and fruits.
- **Water purification and rice cooking**: When between 50 and 60 g of bamboo charcoal is put in one litre of water, the mineral content increases resulting in softer water and improved taste. When cooking with rice, bamboo charcoal will emit infrared radiation and release minerals, which will provide extra sustenance and flavour.
- **Fish tanks and gardening**: Whenever a piece of bamboo charcoal or activated carbon is placed in the tanks of fish, it keeps the water cleaner and makes if healthier for the fishes. When used for soil enrichment in gardening, plants will grow well with better exposure to air and improved humidity conditions.

5.7.5 Bamboo Vinegar

Bamboo vinegar is a by-product of bamboo-charcoal production, which, when heated in the absence of air above a fixed temperature, carbonises, emits volatile. These are condensed to yield vinegar as a by-product (Kabir et al. 2010). It is used for many purposes, from an anti-bacterial agent and agricultural insecticide to deodorants, bath additives, and products to enhance health and beauty. At the time of pyrolysis, bamboo vinegar and bamboo gas are the by-products, along with the charcoal. The escaping smoke condenses in the vents of the oven and drops into a receptacle. The dark-brown condensate smells smoky and separates into a light-yellow layer (bamboo vinegar) and a dark bottom layer (bamboo tar). Depending on the temperature and type of bamboo, the condensed vinegar contains more than 200 organic substances, such as saturated and unsaturated acids, alcohols, aldehydes, and polyphenols. Since bamboo vinegar has beneficial medicinal properties, it is used as an antibacterial, biological preservative (Jiang 2004).

The highest percentage of vinegar acid is produced at an extraction temperature of 300 °C, with an acid content of 8.7% and a pH of 1.8. Because of this low pH, vinegar is applied to acidify the alkaline soil. Charcoal for vinegar has been used in medical products. Bamboo vinegar is used for skin infections and as an antimicrobial, such as in bamboo vinegar spray or soap and also in the cosmetic industry. A bamboo vinegar, contained in "Vital Patch", is said to eliminate unwanted substances from the skin.

Also, the growth of plants is promoted by the use of bamboo vinegar, i.e. as fertiliser, especially for alkaline soils. Thus, plants are protected from damaging microorganisms (Mu et al. 2004).

During pyrolysis of bamboo, about 7% gas is produced, most importantly carbon dioxide, carbon monoxide, methane, ethylene, and hydrogen. This mixture of gases is used directly as fuel. At the end of the charcoal making process, about 2–4% bamboo ash remains, depending upon the species of bamboo and the carbonisation method used. The ash contains minerals such as silica, calcium, potassium, magnesium, sodium, iron, and manganese. They serve agriculture by protecting plants, acidifying the soil, and as fertiliser (Jiang 2007).

5.7.6 Bamboo as Food and Medicine

Bamboo shoots have a high nutrient value, consisting of fibres, proteins, vitamins, minerals, carbohydrates, amino acid, and low fat, giving it very wide potential as food and medicine (Shi and Yang 1992). They are good healthy foods that provide many health benefits to the consumers, especially in Asian and African countries such as China, Indian, Tanzania, Uganda, Ghana and many others where it forms an integral part of several traditional cusines of the people (Bao 2006). Table 5.2 outlines some edible bamboo species and their respective quality. The existence of phytosterols in young bamboo shoots sustains a youthful feeling, healthy dynamism, and longevity to regular consumers as medicine. Freshly harvested shoots contain a significant amount of vitamin A, vitamin B6, vitamin E, niacin, and thiamine (Visuphaka 1985; Xia 1989). They are a rich source of dietary fibre and contain little cholesterols and phytosterols, making it an acceptable natural health food. Medicinally, bamboo shoots contain arginine histidine, leucine, and tyrosine amino acids. The presence of tyrosine in the shoots eases the biochemical metabolism of the body since it is the chief component in adrenals, being the antecedent of adrenaline, which is required to stimulate metabolism in human bodies. Tyrosine also plays a significant part in the function of the thyroid and pituitary glands, which are involved in regulating and producing hormones in the human body. High volumes of fibre and phytosterols in bamboo shoots, reduce cholesterol and fat levels in the blood, making it the most desired health food among patients with lifestyle-related disorders. Bamboo shoots medicinally are used to control the early phases of cancer infection (Kalita and Dutta 2012), since lignans and phytosterols in the shoots possess anticancer properties. The presence of phytosterols in bamboo

Table 5.2 List of some species of bamboo that are edible

Name of species	Quality	Name of species	Quality
Acidosasa edulis	Delicious	*Fargesia robusta*	Edible
Acidosasa lingchuanensis	Edible	*Gigantochloa atter*	Good
Bambusa balcooa	Good	*Gigantochloa levis*	Delicious
Bambusa bambos	Edible	*Gigantochloa ligulata*	Good
Bambusa beecheyana	Good	*Gigantochloa nigrociliata*	Good
Bambusa blumeana	Good	*Gigantochloa pruriens*	Good
Bambusa gibboides	Good	*Gigantochloa robusta*	Good
Bambusa polymorpha	Good	*Gigantochloa thoii*	Good
Bambusa tulda	Good	*Guadua sarcocarpa*	Good
Bambusa tuldoides	Good	*Himalayacalamus falconeri*	Good
Bambusa vulgaris	Edible	*Nastus elatus*	Edible
Chimonobambusa communis	Good	*Oxytenanthera abyssinica*	Edible
Chimonobambusa marcrophylla	Delicious	*Phyilostachys concava*	Edible
Chimonobambusa marmoreal	Delicious	*Phyilostachys incarnate*	Delicious
Chimonobambusa pachystachys	Delicious	*Phyilostachys sulphurea* f. viridis	Good
Chimonobambusa puberula	Delicious	*Phyilostachys acuta*	Delicious
Chimonobambusa quadrangularis	Delicious	*Phyilostachys angusta*	Edible
Chimonobambusa rigidula	Delicious	*Phyilostachys arcana*	Edible
Chimonobambusa szechuanensis	Delicious	*Phyilostachys atrovaginata*	Edible
Chimonobambusa tumidissinida	Delicious	*Phyilostachys bambusoides*	Bitter
Chimonobambusa utilis	Good	*Phyllostachys bambusoides* f. shouzhu	Edible
Chimonocalamus delicatus	Delicious	*Phyllostachys bissetii*	Edible
Dendrocalamus asper	Good	*Phyllostachys circumpilis*	Edible
Dendrocalamus brandisii	Good	*Phyllostachys decora*	Edible
Dendrocalamus giganteus	Good	*Phyllostachys dulcis*	Delicious
Dendrocalamus latiflorus	Good	*Phyllostachys edulis*	Good
Dendrocalamus latiflorus 'Mei-Nung.'	Good	*Phyllostachys edulis* f. edulis	Delicious
Dendrocalamus membranaceus	Edible	*Phyllostachys elegans*	Delicious
Dendrocalamus strictus	Edible	*Phyllostachys erecta*	Edible
Phyllostachys fimbriata	Edible	*Phyllostachys praecox*	Delicious
Phyllostachys fimbriligula	Delicious	*Phyllostachys praecox* f. notata	Edible
Phyllostachys flexuosa	Delicious	*Phyllostachys praecox* f. viridisulcata	Delicious
Phyllostachys glabrata	Delicious	*Phyllostachys prominens*	Good

(continued)

Table 5.2 (continued)

Name of species	Quality	Name of species	Quality
Phyllostachys glauca	Good	*Phyllostachys propinqua*	Good
Phyllostachys glauca f. yunzhu	Edible	*Phyllostachys propinqua* f. lanuginosa	Delicious
Phyllostachys glauca var. variabilis	Edible	*Phyllostachys purpurata* 'Solidstem.'	Edible
Phyllostachys heteroclada	Edible	*Phyllostachys rivalis*	Delicious
Phyllostachys iridescens	Delicious	*Phyllostachys robustiramea*	Edible
Phyllostachys makinoi	Edible	*Phyllostachys rubella*	Edible
Phyllostachys heteroclada	Edible	*Phyllostachys rubromarginata*	Edible
Phyllostachys iridescens	Delicious	*Phyllostachys rutile*	Edible
Phyllostachys makinoi	Edible	*Phyllostachys sapida*	Edible
Phyllostachys meyeri	Edible	*Phyllostachys sulphurea* f. laqueata	Good
Phyllostachys mirabilis	Edible	*Phyllostachys tianmuensis*	Edible
Phyllostachys nidularia	Delicious	*Phyllostachys viridiglaucescens*	Edible
Phyllostachys nidularia f. farcta	Edible	*Phyllostachys vivax*	Delicious
Phyllostachys nidularia f. mirabilis	Edible	*Phyllostachys vivax* 'Huangwenzhu.'	Edible
Phyllostachys nidularia f. speciose	Edible	*Phyllostachys vivax* f. aureocaulis	Edible
Phyllostachys nidularia f. sulfurea	Edible	*Phyllostachys yunhoensis*	Delicious
Phyllostachys nigella	Delicious	*Pleioblastus hindsii*	Edible
Phyllostachys nigra f. henonis	Delicious	*Sasa kurilensis*	Good
Phyllostachys nuda	Delicious	*Sasaella masamuneana*	Edible
Phyllostachys nuda 'Ink-finger.'	Delicious	*Thamnocalamus aristatus*	Edible
Phyllostachys parvffolia	Delicious	*Thyrsostachys siamensis*	Good
Phyllostachys pingyangensis	Edible	*Yushania maling*	Good
Phyllostachys platyglossa	Delicious		

Source Edible Bamboo (20/07/2016)

shoots inhibits the production of cancer-cell growth, carcinogens, cell invasion, and metastasis (Meric et al. 2006). Bamboo leaves, roots, shoots, and stems are used as medicine and food. It can also be used for cosmetic products as well as containers. A good amount of phytosterol can be extracted from freshly harvested bamboo shoots, and this can effectively be utilised for the production of steroidal drugs (Nongdam and Tikendra 2014). The use of bamboo for the production of steroidal drugs will minimise the pressure on Dioscorea and Solanum as the primary natural source of steroidal drugs. The nutrient and medicinal content of shoots of bamboo species are intensely researched.

5.7.6.1 Amino Acids

These nutrients are found to be rich and abundant in most bamboo shoots. Of approximately seventeen kinds of amino acids discovered in bamboo shoots, eight are very useful to the human body (Qiu 1992). A study conducted on shoots of a 300 day-old Bambusa tulda to determine the content of individual amino acids indicated a decrease in the individual amino acids (Giri and Janmejay 2000). Sharma et al. (2004) conducted a study on the shoots to detect certain amino acids in four species of Dendrocalamus and five species of Bambusa that revealed extremely inconsistent results across the species. A study conducted by Nirmala et al. (2008) on freshly harvested, fermented, and canned bamboo shoots of Dendrocalamus giganteus in measured the amount of amino acid content in the species. The results showed that there was a decrease in some amino acids in canned shoots which contain 1.980/100 g fresh weight, and in fermented shoots which contain 2.005/100 g fresh weight, compared to freshly collected young shoots in which the amino acid content was found to be 3.863/100 g fresh weight. In another study to determine the number of amino acids by Nirmala et al. (2007) involving a ten-day-old harvested bamboo shoots of species such as Bambusa bamboos, Bambusa tulda, Dendrocalamus asper, Dendrocalamus giganteus, and Dendrocalamus hamiltonii, shows a substantial reduction in free amino acids compared to freshly harvested shoots. Zhang et al. (2011) in a study on the Phyllostachys praecox shoots of bamboo discovered twelve free amino acids. Of these twelve, six were paramount while the remaining six were found to be supplementary. The six critical amino acids in the Phyllostachys shoots include isoleucine, leucine, lysine, methionine, and valine, while the extra amino acids are alanine, aspartate, glutamine, glycine, histidine, and tyrosine. Because of the loss of amino acids to fluids during heat treatment, there was a reduction in the total free essential amino acid (TFEAA) and total free amino acid (TFAA) in the boiled water of 38.85 and 38.35% respectively. The critical amino acids represented 12–48% of the total amino acid content of bamboo shoots evaluated by Xu et al. (2005) on nine bamboo species from China using conventional methods. Some amino acids in a Dendrocalamus latiflorus shoot exhibited different salt concentrations during a preserving process (Chen et al. 2013), and showed a high rate of reduction in the shoots preserved in a 8% salt concentration, from 16.35/100 g to 6.898/100 g fresh weight. According to a study conducted by Kozukuen et al. (1983), tyrosine was found to be the most abundant total free amino acids in young shoots of Phyllostachys pubescens bamboo.

5.7.6.2 Protein

Bamboo sprouts possess a high amount of protein in the range of 1.49/100–4.04/100 g new weight in fresh bamboo shoots (Chongtham et al. 2011; Sundriyal and Sundriyal 2001). A study conducted on nine bamboo species by (Sharma et al. 2004) concluded that there are variations of proteins in various species. There was a considerable decrease of protein in Bambusa bamboos shoots that had been boiled in a concentrated salt solution. The protein content of the bamboo shoots after 25 min of boiling in a concentration of 10% NaCl solution was low (Pandey and Ojha 2011). A study by Nirmala et al. (2008) on Dendrocalamus giganteus species of bamboo found that canned shoots have the lowest protein content of 1.980/100 g fresh weight compared to fermented shoots which have (2.2005/100 g fresh weight), ten-day-old emerged sprouts have 2.230/100 g fresh weight, and freshly harvested sprouts have (3.863/100 g fresh weight) of Dendrocalamus giganteus.

Devi and Singh (1986) conducted a study and reported an improvement of protein in fermented bamboo sprouts from 3.1 to 7.1% and 8.1% on the third and fifth days, respectively, in the process of fermentation. Zhang et al. (2011) conducted a study on Dendrocalamus asper and Dendrocalamus strictus and found that the amount of protein reduces in bamboo shoots over time. Hence, the amount of protein in eight-day-old Dendrocalamus asper and Dendrocalamus strictus sprouts is 1.21/100 g and 1.91/100 g fresh weight, respectively. The amount of protein was observed to have reduced significantly in Dendrocalamus asper shoots after an additional ten days, namely to 0.86/100 g fresh weight. The amount of protein detected in freshly harvested sprouts of species such as Bambusa bamboo, Bambusa tulda, Dendrocalamus asper, Dendrocalamus strictus, Dendrocalamus giganteus, and Dendrocalamus hamiltonii in an experimental study indicated a reduction in ten-day-old bamboo sprouts as compared to freshly harvested sprouts (Nirmala et al. 2007).

5.7.6.3 Minerals

Bamboo sprouts are full of rich and valuable minerals which include calcium, iron, magnesium, phosphorus, potassium, and sodium. Several metabolic processes in the human body need these minerals to function efficiently. A study by Bhargava et al. (1996) on Bambusa arundinaria and Melocanna baccifera found that there is a high amount of potassium, namely 1400 mg/100 g fresh weight in the former, and a minimum quantity of 20 mg/100 g fresh weight in the latter. In another study by Bhatt et al. (2005), it was reported that there were low quantities of potassium, namely 20 mg/100 g fresh weight in Bambusa tulda, Dendrocalamus hookeri, Dendrocalamus longispathus, and Dendrocalamus sikkimensis. A study conducted by Nirmala et al. (2007) detected minor decreases in potassium content in ten-day-old harvested bamboo species shoots which includes Bambusa bamboo, Bambusa tulda, Dendrocalamus asper, and Dendrocalamus giganteus. It was observed that, with the exception of Dendrocalamus hamiltonii, potassium levels dropped from 416 mg/100 g to 210 mg/100 g fresh weight in the ten-day-old sprouts, which is about half of the fresh amount. A study by Feleke (2013) was conducted on Oxytenanthera abyssinica bamboo sprouts in Ethiopia to determine the amount of potassium, and high quantities of potassium were found depending on the location of the harvest. Sprouts harvested from the Assosa region contain

6.63%, sprouts from Dhidhessa 7.02%, and sprouts from Pawa is 7.15%. Comparatively, the amount of potassium in this species (Oxytenanthera abyssinica) is higher than in popular vegetables such as amaranthus spinous, hibiscus, and Solanum marcrocarpon.

Pandey and Ojha (2011), conducted a study on Bambusa tulda, Dendrocalamus asper, and Dendrocalamus strictus, to determine the variations of mineral content in various bamboo species by exposing them to different salt concentrations over various time durations. They found that calcium, phosphorus, magnesium, potassium, and sodium in the fresh bamboo sprouts showed no substantial changes between the treatments. Instead, the amount of sodium improved significantly from 0.07 mg/100 g fresh weight in the fresh sprouts to 1.15 mg/100 g fresh weight in the sprouts boiled in a 10% NaCl for 25 min. There were traces of minor discrepancies in the levels of elements in the freshly harvested sprouts, fermented, and non-salted canned species of Dendrocalamus giganteus, including cadmium, cobalt, manganese, nickel, and selenium. According to a study conducted by Nirmala et al. (2008), the amount of zinc decreased in fermented and canned bamboo sprouts as compared to fresh sprouts. Medically, it is suggested that a daily dose of potassium for humans should range between 2.0 and 5.5/day (Belitz and Grosch 1999) because it enhances and protects the heart by sustaining normal BP and a steady heartbeat. Women during pregnancy and infant nursing require high amounts of iron (Tapiero et al. 2001), and consistent intake of iron rich bamboo sprouts can boost this intake.

5.7.6.4 Carbohydrate

The percentages of carbohydrates in bamboo sprouts, e.g. the highly edible Bambusa nutans, Bambusa vulgaris, Dendrocalamus aspers and Dendrocalamus strictus, are 3.3, 3.4, 2.9, and 0.6%, respectively (Kumbhare and Bhargava 2007). It was also found that heat treatment through boiling does increase the amount of carbohydrate in bamboo sprouts as a result of the hydrolysis of complex polysaccharides into single monosaccharide units of sugar. In a study conducted by Pandey and Ojha (2011), it was found that some carbohydrates decline when the sprouts of Bambusa bamboos, Bambusa tulda, Bambusa strictus and Dendrocalamus asper are boiled in various different concentrations of saline solutions. The existence of salt in the solution might have an influence on decreasing the carbohydrate content by enhancing the hydrolysis of carbohydrate in the process of boiling. In a study conducted by Pandey and Ojha (2013) on the rise of carbohydrate levels with an increase in the age of the sprouts, it was found that the amount of carbohydrate in a two-day-old sprout of bamboo increases from 1.45/100 g to 2.46/100 g fresh weight after sixteen days. In a study by Nirmala et al. (2007) on the decrease of carbohydrate with an increase in the age of bamboo, a ten-day-old sprout contains 2.30/100 g fresh weight carbohydrates which is less than a freshly harvested sprout which contains 5.42/100 g fresh weight. The carbohydrate content in a fermented sprout is

1.504/100 g fresh weight, which means that it has declined when compared to the content in a fresh sprout of Dendrocalamus giganteus which is 5.103/100 g fresh weight (Nirmala et al. 2008). In other findings by (Devi and Singh 1986), carbohydrate content also decreased in the fermentation of Phyllostachys humilis sprouts.

5.7.6.5 Fibre

The high content of dietary fibres and low fat in the shoots help to control blood pressure and the thickening of arteries, hypertension, obesity, cardiovascular ailments, thus protecting the human body from coronary diseases and possible carcinogens (Kalita and Dutta 2012). The consumption of a fibre-rich bamboo diet contributes to a decrease in undesirable cholesterols (low-density and very low-density lipoprotein) in the blood, lowers insulin demand, keeps the digestive tracks healthy, and improves laxative properties (Howarth et al. 2001; Park et al. 2005). Bhatt et al. (2005) and Park and John (2009) claim that eating bamboo shoots continuously enhances bowel movements and lipid profiles in young women. Bhatt et al. (2005) found maximum occurrences of crude fibre (35.5%) in Melocanna baccifera species bamboo shoots. Sharma et al. (2004) also detected a differences in the total fibre, cellulose, hemicellulose, and lignin content in bamboo sprouts of Bambusa arundinaria, Bambusa polymorpha, Bambusa tulda, Dendrocalamus calostachyus, Dendrocalamus giganteus, Dendrocalamus membranaceus, and Dendrocalamus strictus. According to Nirmala et al. (2007), the quantity of dietary fibre increased to twice the amount in ten-day-old sprouts compared to fresh sprouts in Bambusa bamboos, Bambusa tulda, Dendrocalamus asper, Dendrocalamus giganteus, and Dendrocalamus hamiltonii. An investigation conducted by Kumbhare and Bhargava (2007) revealed no substantial reduction in fibre content after boiling bamboo sprouts. However, dietary components such as ADF, NDF, cellulose, hemicellulose, and lignin in fermented sprouts of Dendrocalamus gigantic increased as compared to freshly harvested sprouts (Chongtham et al. 2011).

5.7.6.6 Phytosterol

Phytosterol is similar in structure to cholesterol and is found in most plants, and its existence in fresh or fermented bamboo sprouts is important (Miettinen and Gylling 2003). Several researchers have studied the importance of phytosterol in bamboo sprouts in contributing to its life sustaining qualities (Kritchevsky and Chen 2005; Phillips et al. 2005; Ostlund 2007). Using ultra performance liquid chromatography, the phytosterol content in the sprouts of Pleioblastus amarus, Pleioblastus pubescens, Dendrocalamus latiflorus and Pleioblastus praecox were evaluated by Lu et al. (2009), and they detected higher levels of β-sitosterol as compared to other sterols like campesterol and stigmasterol. According to Choi et al. (2007), Bradford and Awad (2007), and Woyengo et al. (2009), phytosterol-rich diets help to decrease breast, colon, and prostate cancer; the development of tumour growth is subdued by interference in the apoptosis, cell cycle, and tumour metastasis. The existence of phytosterol also hinders the absorption of dietary cholesterol and cholesterol esterification in intestinal mucosa by decreasing of cholesterol levels in the blood. The main sterols available in bamboo sprouts include β-sitosterol, campesterol, and stigmasterol, according to Lachance and He (1998). A study by Sarangthem and Singh (2003) found that that sitosterol is the major phytosterol in bamboo sprouts. A sizeable quantity of phytosterol are extracted from freshly harvested bamboo sprouts and effectively used to produce steroidal drugs which help to reduce the demand on the Dioscorea and Solanum species as the main natural source of steroidal drug components.

5.7.6.7 Phenols

Phenols are essential bioactive compounds in plants that exhibit strong natural antioxidative and anti-inflammatory properties and are sometimes antimicrobial (Rotelli et al. 2003; Puupponen-Pimia et al 2005; Lehane and Saliba 2008; Oboh and Ademosun 2012). According to Oboh and Ademosun (2012), Rice-Evans et al. (1997), and Lu et al. (2005), phenolic compounds in bamboo leaves possess antioxidant qualities. In a study conducted by Velioglu et al. (1998) on the link between the antioxidant properties of a plant and its phenolic compound content, it was found to be statistically relevant. According to Zhang et al. (2011), the quantity of phenolic compounds changes in fresh bamboo sprouts boiled, steamed, and stir-fried, as there is a slight decrease in the overall phenolic content in both boiled and stir-fried sprouts as compared to fresh ones. However, in steamed shoots, there is an increase in the level of phenolic compounds. The reduction in boiled or stir fried sprouts may be due to the decomposition of the phenolic compounds during the heat-treatment process. Yamaguchi et al. (2003) opine that heat treatment might lead to the inactivation of existing phenolic oxidases which are responsible for the decomposition of polyphenols. The higher level of overall phenolic compounds and the increase in antioxidant action are interrelated (Baardseth et al. 2010). A study by Pandey and Ojha (2013) on Bambusa tulda, Dendrocalamus asper, and Dendrocalamus strictus reported variations in phenolic compounds at different optimum harvesting time of the shoots. The concentration of vanillic acid was reduced in Bambusa tulda, but concentration increased in both Dendrocalamus asper from 0.009/100 g to 1.262/100 g fresh weight and Dendrocalamus strictus from 0.273/100 g to 2.563/100 g fresh weight. In another study by (Pandey and Ojha 2011) on some bamboo species sprouts boiled in a higher salt concentration for an extensive period, a reduction in an overall phenolic content in the sprouts was indicated, including Bambusa bamboos, Bambusa tulda, and Dendrocalamus asper.

5.7.6.8 Taxiphyllin in Bamboo Shoots

Bamboo shoots comprise a range of quantities of cyanogen glycosides known as taxiphyllin (Nartey 1980; Hunter and Fenge 2000; Vetter 2000). The β-glycosidase, when released in the tissues of a disturbed bamboo sprout, acts on taxiphyllin to produce hydrogen cyanide which is harmful, and its toxicity should not surpass the toxic level in humans (Seigler 1991). The hydrogen cyanide in bamboo is harmful for humans and requires further treatment to ensure complete removal through boiling (EFSA 2004). Earlier studies to determine the cyanogen content in some bamboo species were conducted by Schwarzmaier (1977) and Hunter and Fenge (2000). Jones (1998) investigated on the content of cyanide (HCN) in various sections of the bamboo plants of different species, and the results show that there are differences in content. Moreover, the quantity of cyanogen glycoside in bamboo sprouts of most edible species has a maximum concentration centered in the sprout tip. Investigations conducted by Haque and Bradbury (2002) concluded that the overall cyanide content in Bambusa arundinacea sprout (1010 ppm) was much higher than in other plant parts like an apricot stone (785 ppm) and sorghum leaf (750 ppm). Edible bamboo sprouts with high toxic levels of cyanogenic compounds can be completely detoxified by boiling in water for about two hours. When bamboo shoots are improperly treated before consumption by humans, the following symptoms may be experienced: a drop-in blood pressure, dizziness, headache, rapid respiration, vomiting, convulsion, coma, and stomach pains (FSANZ, 2004). Hydrogen cyanide interferes with the appropriate performance of cytochrome oxidase preventing normal cellular respiration. The existence of HCN in a concentration of 0.5–3.5 mg/Kg body weight, according to FSANZ findings, may cause serious health problem that can result in death. An HCN is a respiratory poison found in high concentrations in bamboo shoots and can cause serious health concerns if consumed untreated. If bamboo sprouts are not well processed before consumption, stomach ailments and other mild health-related problems may also occur. The processing of bamboo sprouts for consumption is based on traditional Asian methods. However, these processing methods are crude and lack a scientific foundation. In recent years, scientific processing methods have been adopted to decrease cyanogenic content and also to evaluate the toxicity of processed bamboo foods. The methods applied to reduce cyanide and also to remove bitterness from the fresh or fermented sprouts includes boiling the sprouts at 100 °C for 48 h, which successfully eliminates 97% of the cyanide (Ferreira et al. 1995). Tender sprouts can be cut into thin slices, dried, boiled in salt solutions, and drained off to reduce cyanide content. The process of steaming bamboo sprouts significantly reduces hydrocyanic acid content (Tripathi 1998).

In an investigation by Pandey and Ojha (2011), they detected that bamboo sprouts boiled in water with various different concentrations of NaCl at several time intervals produced the greatest cyanide reduction with the least nutrient loss. The cyanogen in Bambusa tulda and Dendrocalamus asper sprouts significantly dropped without much decrease in the content of carbohydrate, phosphorus, potassium, protein, sodium, and total phenol when the sprouts were boiled for 25 min in 1, 5, and 10% NaCl solutions. A study by Wongsakpairod (2000) revealed that HCN could be eliminated when the sprouts of harvested bamboo are exposed to superheated steam because taxiphyllin decomposes at a temperature of over 116 °C.

The addition of banana leaves by the Adi tribe of Arunachal Pradesh was reported by Bhardwaj et al. (2005) to be effective way to remove toxic cyanogen during bamboo sprout fermentation. The process involves constraining bamboo sprouts under big stones and leaving them for 3–4 months to lower the bitter taste of fermentation. According to Choudhury et al. (2012), formulating appropriate processing methods for the production of healthy sprouts is essential to yield products with longer life spans.

5.7.7 Hydraulic Application of Bamboo

Bamboo culms have been used in the hydraulic industry as pipes to safely and continuously supply water to a large rural population in Tanzanian. A bamboo culm for piping should have diameters in the ranges of 3.5, 5.0, 9.0, 10.0, and 12.5 cm with a 4.0 m length, having an internal water pressure up to 6 atmospheres without leaking at the joints (Lipangile 1991). This system is employed to transport water through a network of 150 km of bamboo pipelines in Tanzania and as gutters for rainwater harvesting (Lipangile 1991). In a study, Singh (1979) reported and discussed an original method of drip irrigation that is made out of split bamboo and has been practised in Meghalaya State in India for a long time. The method is simple and of valuable scientific importance and is extensively used on steep slopes of up to 10%. This type of well is drilled down to a depth of 50–80 ft. Also, Baqui and Angeles (1992) evaluated the economics of bamboo-drip irrigation systems and reported that the low-level technology used in its fabrication, installation, operation, and maintenance encourages adoption by farmers. The use of bamboo drip for making inexpensive irrigation and water-supply wells is a promising technology in some developing countries in Asia and Africa. The cost of the bamboo screen is only 20% of the expense of a conventional agricultural screen made of steel and coated with copper. The curing process of bamboo culms does not require chemicals but rather air drying, heat treatment, or submerging in water for about 90 days. The use of bamboo culms for pumping water is safe for domestic purposes. The system, according to a study conducted by the authors, lasted for a little over ten years based on a survey of fifteen farmers. These wells provide an excellent example of simple technology well suited to the needs of small-sized land holdings.

5.7.8 Soil Conditioning

Many countries in Asia (China and India, among others) use bamboo propagation as a tool for soil-nutrient reclamation and rehabilitation of degraded mined lands.

Bamboo is suitable for such rehabilitation because it binds loose soils to prevent erosion and is very versatile, growing in a variety of soils that are poor in mineral and nutrient content (Christianty et al. 1997).

5.7.9 Transportation and Navigation

Bamboo is used to make components such as: yokes, vehicle shafts, and rollers for moving heavy objects employed in land transportation. Masts and spars for boats, or lifts, boat poles, seats and false bottoms, and ribs for boat canopies are examples of bamboo products used in navigation.

References

Abdul Khalil HPS, Bhat IUH, Jawaid M, Hermawan AZ (2012) Bamboo fibre reinforced biocomposites: a review. Mater Des 42:353–368

Amanda S, Untao S (2001) Fracture properties of bamboo. Compos B 32:451–459

Baardseth P, Bjerke F, Martinsen BK, Skrede G (2010) Vitamin C, total phenolics and antioxidative activity in tip-cut green beans (Phaseolus vulgaris) and Swede rods (Brassica napus var. napobrassica) processed by methods used in catering. J Sci Food Agric 90(7):1245–1255

Bansal AK, Zoolagud SS (2002) Bamboo composites: material of the furniture. J Bamboo Rattan 1 (2):119–130

Bao J (2006) The nutrition and bioactive function of bamboo shoots. Food Nutr China 4:2–3

Baqui MA, Angeles HL (1992) Economics of bamboo drip irrigation system. Agric Eng J 1 (4):165–172

Belitz HD, Grosch W (1999) Food chemistry, 2nd edn. Springer, New York

Bhardwaj R, Singh RK, Wangchu L, Sureia A K (2005) Bamboo shoots consumption: traditional wisdom and cultural invasion. In: Proceedings of the national conference on Arunachal Pradesh: tradition in transition, link ecology, economics and ethics, NERIST, Nirjuli

Bhargava A, Kumbhare V, Srivastava A, Sahai A (1996) Bamboo parts and seeds for an additional source of nutrition. J Food Sci Technol 33(2):145–146

Bhatt BP, Singh K, Singh A (2005) Nutritional values of some commercial edible bamboo species of the North Eastern Himalayan region. India, J Bamboo Rattan 4(2):111–124

Bradford PG, Awad AB (2007) Phytosterols as anticancer compounds. Mol Nutr Food Res 51 (2):161–170

Chen G, Wang L, Zheng ZJ (2013) Effect of salt concentration on quality of bamboo shoots during pickling. Food Sci 34(15):48–52

Choi J, Lee E, Lee H (2007) Identification of campesterol from chrysanthemum coronarium l. and its antiangiogenic activities. Phytotherapy Res 21(10):954–959

Chongtham N, Bisht MS, Haorongbam S (2011) Nutritional properties of bamboo shoots potential and prospects for utilisation as health food. Compr Rev Food Sci Food Saf 10(3):153–168

Choudhury D, Jatindra SK, Sharma GD (2012) Bamboo shoot: microbiology, biochemistry and technology of fermentation—a review. Indian J Tradit Knowl 11(2):242–249

Christanty L, Kimmins JP, Mailly D (1997) Without bamboo, the land dies': a conceptual model of the biogeochemical role of bamboo in an Indonesian agroforestry system. Ecol Manage 91 (1):83–91

Corradi S, Isidori T, Corradi M, Soleri F, Olivari L (2009) Composite boat hulls with bamboo natural fibres. Int J Mater Prod Tec 36:73–89
Devi SP, Singh HT (1986) Studies on the chemical and nutritional changes of bamboo shoots during fermentation. J Food Sci Technol 23(6):338–339
Dransfield S, Eijada WA (1995) Bamboos, plant resources of South East Asia Network, Nr. 7. PROSEA and Backhuys, Bogor, Indonesia
Durst S (2009) The panda food in your pants. Business 2.0.7(9)
EFSA (2004) Opinion of the scientific panel on food additives flavourings, processing aids and materials in contact with food (AFC) on hydrocyanic acids in flavourings and other food ingredients with flavouring properties. Eur Food Saf Authority J 105:1–28
Feleke S (2013) Site factor on the nutritional content of arundinaria alpine and oxytenanthera abyssinica bamboo shoots in Ethiopia. J Hortic For 5(9):115–121
Ferreira VLP, Yotsuyanagi K, Carvalho CRL (1995) Elimination of cyanogenic compounds from bamboo shoots dendrocalamus giganteus munro. Trop Sci 35(4):342–346
FSANZ (2004) Cyanogenic glycosides in cassava and bamboo shoots: a human health risk assessment. Tech, Rep 28
Fu J (2001) Chinese moso bamboo: its importance. ABS Mag 22(5)
Fu JH (2003) Bamboo charcoal products. Unpublished report, China
Ghana Bamboo Initiatives (2016) http://ghanabamboobikes.org. Accessed 13 June 2016
Giri SS, Janmejay LS (2000) Effect of bamboo shoots fermentation and ageing on nutritional and sensory qualities of Soibum. J Food Sci Technol 37(4):423–426
Guadua Bamboo (2016) https://www.guaduabamboo.com/uses/corrugated-bamboo-roofing-sheets. Accessed 20 July 2016
Guan M (2004) Manual for bamboo charcoal production and utilization, bamboo engineering research center. East Nanjing Forestry University, Nanjing
Haque MR, Bradbury JH (2002) Total cyanide determination of plants and foods using the picrate and acid hydrolysis methods. Food Chem 77(1):107–114
Hongyan C, Miao M, Ding X (2009) Influence of moisture absorption on the interfacial strength of bamboo/ vinyl ester composites. Compos A 40:2013–2019
Howarth NC, Saltzman E, Roberts SB (2001) Dietary fibre and weight regulation. Nutr Rev 59 (5):129–139
Hunter I, Feng'e Y (2000). Cyanide in Bamboo Shoots. http://www.inbar.int/publication/txt/ INBAR_Working_Paper_No39.htm. Accessed 12 Oct 2016
Ip AWM, Barford JP, McKay G (2008) Production and comparison of high surface area bamboo-derived active carbons. Bioresour Technol 99:8909–8916
Ismail H, Shuhelmy S, Edyham MR (2002) The effects of a silane coupling agent on curing characteristic and mechanical properties of bamboo fibre filled natural rubber composites. Eur Polym J 38:39–47
Ismail H, Edyham MR, Wirjosentono B (2001) Dynamic properties and swelling behaviour of bamboo filled natural rubber composites: the effect of bonding agent. Iran Polym J 10(6):377–383
Janssen JJA (2000) Designing and building with bamboo. International network for bamboo and rattan technical report no. 29, p 207
Jiang S (2004) Training manual of bamboo charcoal for producers and consumers. Bamboo Engineering Research Center, Nanjing Forestry University, Nanjing
Jiang Z (2007) Bamboo pulp and paper making (Chapter 11); bamboo charcoal. Vinegar and active carbon (Chapter 13). In: Bamboo and rattan in the world. China forestry publishing house, Beijing, pp 178–198, 214–226
Jones D (1998) Why are so many food plants cyanogenic? Phytochem 47(2):155–162
János Vetter, (2000) Plant cyanogenic glycosides. Toxicon 38(1):11–36
Kabir E, Kim KH, Ahn JW, Hong OF, Sohn JR (2010) Barbecue charcoal combustion as a potential source of aromatic volatile organic compounds and carbonyls. J Hazard Mater 174:492–499
Kalita T, Dutta U (2012) A comparative study on the indigenous usage of bamboo shoot in the health care practices in NE India. The Clarion 1(2):130–141

Khanam PN, Abdul HPS, Khalil G, Ramachandra R, Naidu SV (2011) Tensile, flexural and chemical resistance properties of sisal fibre reinforced polymer composites: effect of fibre surface treatment. Polym Environ 19:115–119

Kozukuen E, Kozukue N, Kurosaki T (1983) Organic acid, sugar and amino acid composition of bamboo shoots. J Food Sci 48(3):935–938

Kritchevsky D, Chen SC (2005) Phytosterols-health benefits and potential concerns: a review. Nutr Res 25(5):413–428

Kumbhare V, Bhargava A (2007) Effect of processing on the nutritional value of central Indian bamboo shoots. Part I, J of Food Sci 44(3):935–936

Lachance PA, He YH (1998) Hypocholesterolemic composition from bamboo shoots. PCT International Patent PCT/US98/12556

Lalhruaitluanga H, Jayaram K, Prasad MNV, Kumar KK (2010) Lead(II) adsorption from aqueous solutions by raw and activated charcoals of Melocanna baccifera Roxburgh (bamboo)—a comparative study. J Hazard Mater 175:311–318

Latif A-M, Liese W (1995) Utilization of bamboo. In: Razak A-O, Latif A-M, Liese W, Norini H (eds) Planting and utilization of bamboo in Peninsula Malaysia. FRIM research pamphlet no. 118. Forest Research Institute Malaysia, Kuala Lumpur, Malaysia, pp 50–102

Lehane AM, Saliba KJ (2008) Common dietary flavonoids inhibit the growth of the intraerythrocytic malaria parasite. BMC Research Notes, vol 1, article 26

Lichtenstadt P (1864) Verification of bamboo fibre, US Patent & Trademark Office

Liese W, Silbermann S (2010) Bamboo charcoal: properties and utilization. Mag Am Bamboo Soc 31(6):2–9

Lipangile TN (1991) Manufacture and construction of bamboo water supply systems. J Am Bamboo Soc 8(1, 2):191–198

Lobovikov M, Paudel S, Piazza M, Ren H, Wu J (2007) World bamboo resources: an mathematic study prepared in the framework of the global forest resources assessment (2005). Food and Agriculture Organization (FAO) of the United Nations, Rome

Lu BY, Bao JF, Shan L, Zhang Y (2009) Technology for supercritical CO_2 extraction of bamboo shoots oil and components of the product. Trans Chin Soc Agri Eng 25(8):312–316

Lu B, Wu X, Tie X, Zhang Y (2005) Toxicology and safety of antioxidant of bamboo leave. Part 1: acute and subchronic toxicity studies on antioxidant of bamboo leaves. Food Chem Toxicol 43(5):783–792

Mahdavi M, Clouston PL, Arwade SR (2012) A low-technology approach toward fabrication of laminated bamboo lumber. Constr Build Mater 29:257–262

Malanit P (2009) The suitability of Dendrocalamus asper Backer for oriented strand lumber. Dissertation, University of Hamburg

Meric J, Rottey S, Olaussen K (2006) Cyclooxygenase-2 as a target for anticancer drug development. Crit Rev Oncol/Hematol 59(1):51–64

Miettinen TA (2003) Gylling H (2003) Non-nutritive bioactive constituents of plants: phytosterols. Int J Vitam Nutr Res 73(2):127–134

Minke G (2012) Building with bamboo. Verlag Birkhauser Architecture. ISBN 3034607482, 9783034607483

Mohana Rao K, Ratna Prasad AV (2010) Fabrication and testing of natural fibre composites: vakka, sisal, bamboo, and banana. Mater Des 31(1):4508–4513

Mu J, Uehara T, Furuno T (2004) Effect of bamboo vinegar on regulation of germination and rapid growth of seed plants II: composition of moso bamboo vinegar at different collection temperature and its effects. J Wood Sci 50:470–476

Muller I, Rebelo C (2015) Bamboo Worldwide: The current market and future potential. http://www.ecoplanetbamboo.com/files/bamboo_worldwide.pdf. Accessed 17 Oct 2016

Nartey F (1980) Toxicological aspects of cyanogens in tropical foods. In: Smith RL, Bababummi ER (eds) Toxicology in the tropics. Taylor & Francis, London, pp 53–73

Naxium M (2001) Biodiversity and resources exploitation of Bamboo in China. In: Zhu Z (ed) Sustainable development of bamboo and Rattan sectors in tropical China. Sector proceedings No. 6. INBAR and China Forestry Publishing House

Nirmala C, David E, Sharma ML (2007) Changes in nutrient components during ageing of emerging juvenile bamboo shoots. Int J Food Sci Nutr 58(8):612–618

Nirmala C, Sharma ML, David E (2008) A comparative study of a nutrient component of freshly harvested, fermented and canned bamboo shoots of Dendrocalamus giganteus Munro. J Am Bamboo Soc 21(1):33–39

Nirmal U, Jamilhashim KO (2012) Low adhesive wear and frictional performance of bamboo fibres reinforced epoxy composite. Tribol Int 47:122–133

Nongdam P, Tikendra L (2014) The Nutritional Facts of Bamboo Shoots and Their Usage as Important Traditional Foods of Northeast India. Int Schol Res Not 17

Oboh G, Ademosun AO (2012) Characterization of the antioxidant properties of phenolic extracts from some citrus peels. J Food Sci Technol 49(6):729–736

Ostlund RE (2007) Phytosterols, cholesterol absorption and healthy diets. Lipids 42(1):41–45

Pandey AK, Ojha V (2011) Precooking processing of bamboo shoots for removal of anti-nutrients. J Food Sci Technol 51(1):43–50

Pandey AK, Ojha V (2013) Standardisation of the harvesting age of bamboo shoots on nutritional and anti-nutritional components. J Forest Res 24(1):83–90

Park EJ, John DY (2009) Effects of bamboo shoot consumption on lipid profiles and bowel function in healthy young women. Nutrition 25(7–8):723–728

Park Y, Hunter DJ, Spiegelman D (2005) Dietary fibre intake and risk of colorectal cancer. J Am Med Assoc 294(22):2849–2857

Phillips KM, Ruggio DM, Ashraf-Khorassani M (2005) Phytosterol composition of nuts and seeds commonly consumed in the United States. J Agric Food Chem 53(24):9436–9445

Ratna Prasad AV, Mohana Rao K (2011) Mechanical properties of natural fibre rein- forced polyester composites: jowar, sisal and bamboo. Mater Des 32:4658–4663

Puupponen-Pimia R, Nohynek L, Hartmann-Schmidlin S (2005) Berry phenolics selectively inhibit the growth of intestinal pathogens. J Appl Microbiol 98(4):991–1000

Qiu FG (1992) The recent development of bamboo foods. Proceedings of the international symposium on industrial use of bamboo. International Tropical Timber Organization and Chinese Academy, Beijing, pp 333–337

Rice-Evans CA, Miller NJ, Paganga G (1997) Antioxidant properties of phenolic compounds. Trends Plant Sci 2(4):152–159

Rodie JB (2007) Quality fabric of the month: spotlight on Bamboo. Textile World, March/April 2007

Rotelli AE, Guardia T, Juarez AO, de la Rocha NE, Pelzer LE (2003) Comparative study of flavonoids in experimental models of inflammation. Pharmacol Res 48(6):601–606

Rubio-Luna G (2007) Arte y Man˜as de la Guadua—unaguiasobre el uso productive de unbambu´ gigante. Editor Info Art, Bogotá

Sarangthem K, Singh TN (2003) Transformation of fermented bamboo (Dendrocalamus asper) shoots into phytosterols by microorganisms. J Food Sci Technol 40:622–625

Schwarzmaier U (1977) Cyanogenesis of Dendrocalamus: taxiphyllin. Phytochem 16(10):1599–1600

Seigler DS (1991) Cyanide and a cyanogenic glycoside, in herbivores: their interactions with secondary plant

Sharma ML, Nirmala C, David E (2004) Variations in nutrient and nutritional components of juvenile bamboo shoots. Panjab Univ Res J 54:101–104

Sharma B, Gatóo A, Bock M, Mulligan H, Ramage M (2014) Engineered bamboo: state of the art. Proc Inst Civil Eng. doi:10.1680/coma.14.00020

Shi QT, Yang KS (1992) Study on relationship between nutrients in bamboo shoots and human health. In: Proceedings of the International Symposium on Industrial Use of Bamboo. International Tropical Timber Organization and Chinese Academy, Beijing

Shi S, Walker J, Wood-Based (2006) Composites: plywood and veneer-based products. Primary wood processing ed. USA, pp 391–426

Siriwan K, Suthamnoi W (2009) Physical properties of polyolefin/ bamboo charcoal composites. J Met Mater Miner 19(1):9–15

Singh A (1979) Bamboo Drip Irrigation System. J Agr Eng 16(3):103–107

Sundriyal M, Sundriyal RC (2001) Wild edible plants of the Sikkim Himalaya: nutritive values of selected species. Econ Bot 55(3):377–390

Tan Z, Qiu J, Zeng H, Liu H, Xiang J (2011) Removal of elemental mercury by bamboo charcoal impregnated with H_2O_2. Fuel 90:1471–1475

Tapiero H, Gate L, Tew KD (2001) Iron: deficiencies and requirements. Biomed Pharmacother 55 (6):324–332

Tong J Arnell RD Ren LQ (1998) Dry sliding wear behaviour of bamboo. J Wear 221:37–46

Tistl M, Velásquezgil JA (2001) Roofs instead of tents: a reconstruction project in the Colombian coffee zone after the earthquake of January 25th, 1999. In: Bamboo in disaster avoidance. Proceedings of the international workshop on the role of bamboo in disaster avoidance. 6–8 Aug 2001, Quayaquil, Ecuado

Tripathi YC (1998) Food and nutrition potential of bamboo. MFP News 8(1):10–15

Van der Lugt P, Van den Dobbelsteen AAJF, Janssen JJA (2005) An environmental, economic and practical assessment of bamboo as a building material for supporting structures. Construct Build Mater 20(9):648–656

Velioglu YS, Mazza G, Gao L, Oomah BD (1998) Antioxidant activity and total phenolic in selected fruits, vegetables and grain products. J Agric Food Chem 46(10):4113–4117

Visakh PM, Kristina O, Mathew AP (2012) Crosslinked natural rubber nanocomposites reinforced with cellulose whiskers isolated from bamboo waste: processing and mechanical/thermal properties. Composites: Part A 43:735–741

Visuphaka K (1985) The role of bamboo as a potential food source in Thailand, In: Proceedings of the international bamboo workshop, recent research on bamboos, Hangzhou, China, Oct 1985, pp 301–303

Wang FY, Wang H, Ma JW (2010) Adsorption of cadmium (II) ions from aqueous solution by a new low-cost adsorbent—Bamboo charcoal. J Hazard Mater 177:300–306

Wongsakpairod T (2000) Bamboo shoot drying using superheated steam. MEng. Thesis, King Mongkut's University of Technology, Bangkok

Woyengo TA, Ramprasath VR, Jones PJH (2009) Anticancer effects of phytosterols. Eur J Clin Nutr 63(7):813–820

Xia NH (1989) Analysis of nutritive constituents of bamboo shoots in Guangdong, Acta Botanica Austro Sinica, vol 4, pp 199–206

Xu S, Wan-You CAO, Song QY (2005) Analysis and evaluation of protein and an amino acid nutritional component of different species of bamboo shoots. J Food Sci 26(7):222–227

Yu HP, Shen K, Wang ZT, Mu LL, Li GQ (2011) Population control of the yellow-spined bamboo locust, Ceracris kiangsu, using urine-borne chemical baits in bamboo forest. Entomol Exp Appl 138(1):71–76

Yamaguchi T, Katsuda M, Oda Y (2003) Influence of polyphenol and ascorbate oxidases during cooking process on the radical scavenging activity of vegetables. Food Sci Technol Res 9 (1):79–83

Zhang J, Ji R, Hu Y, Chen J, Ye X (2011) Effect of three cooking methods on nutrient components and antioxidant capacities of bamboo shoot (Phyllostachys praecox C.D. Chu et C.S. Chao). J Zhejiang Univ SCI B. Biomed Biotechnol 12(9):752–759

Chapter 6
Current Trend in Bamboo Analysis

Abstract The focus of most design and manufacturing firms is to improve product quality with high efficiency and at lower cost. Material properties and cost are critical for achieving production efficiency. Finite Element Analysis (FEA) has proven to be a valuable tool for limiting inefficiencies and improving reliability in most engineering designs. A brief review of the FEA process and its use in wood and bamboo characterisation is presented in this chapter. Additionally, the use of FEA for simulation of bamboo behaviour under tensile, compression, and bending is discussed. The second part of this chapter discusses some significant studies conducted on bamboo as a Functionally Graded Material (FGM). Understanding the FGM property of bamboo will legitimise it for many more industrial and engineering use. Other artificial materials can be developed to exhibit similar optimised functionalisation. Experimental work is also carried out to show the microstructure of the vascular bundle. An optimised design that is a function of adaptability to environmental loading is observed.

6.1 Introduction

The traditional way of testing engineering materials to determine their properties has experienced a shift from costly experimentation only to full integration of modeling and simulation to save the cost and time required for testing. Bamboo, due to its unique characteristic of being a grass but ability to grow as tall as trees, must be considered as a Functionally Graded Material to ensure accuracy in its computational analysis. The geometry, fibre distribution, and mechanical properties of bamboo culms in relation to FGM principles are reviewed in this chapter. The use of FEA to model bamboo subjected to tensile, compression, and bending is also presented.

E.T. Akinlabi et al., *Bamboo*, DOI 10.1007/978-3-319-56808-9_6

6.2 Introduction to Finite Element Analysis of Bamboo

Bamboo is a natural composite comprising cellulose fibre and a matrix of lignin. The bamboo culm is made up of 20–30% fibres that are spread out non-uniformly and longitudinally and can be assumed to be an approximately hollow cylindrical shell (Silva et al. 2006). This distribution is observed throughout the entire wall thickness of the culm. Bamboo can, therefore, be considered an orthotropic material because of the fibre distribution and orientation (Ghavami 1995; Lo et al. 2004). The fibre distribution concentration decreases progressively from the exterior to the interior of culm wall (Silva et al. 2006).

The advantages of using bamboo for structural and other engineering purposes can be attributed to factors such as the growth rate (3–4 years to maturity), it's being a cheaper alternative and lightweight, and, unlike steel, requiring no pre-processing (Arun et al. 2002; Verma et al. 2012). The drawbacks can be linked to the cylindrical profile, which inherently presents a stabilisation problem, while the hollow characteristic significantly reduces its ability to resist radial loading. Lastly, it has a weak surface hardness that can compromise durability in abrasive environments (Khatry and Mishra 2012).

6.2.1 Description of the Finite-Element-Analysis Process

The use of finite elements has assumed prominence in analysing physical phenomena and the behaviour of engineering materials such as metals, composites, and wood. FEA involves the discretization of very complex domains made up of a collection of simpler elements of finite sizes. The nature of FEA enables the analysis of global problems using several differential equations that encapsulate the boundary conditions and compatibility between elements (Tankut et al. 2014). The theoretical framework and fundamental studies in the adoption and application of FEA have been presented by researchers (Cook 1981; Tanvir and Utku 1987; Bathe 1996; Zienkiewicz and Taylor 1988; Zienkiewicz and Taylor 1989). There are three main phases required for any FEA, namely, the pre-processing, solution, and post-processing phases. Pre-processing involves problem subdivision by discretizing the created domain solution into finite element forms. A geometrical representation of the physical behaviour of elements is created and serves as an approximate elemental solution. The development of equations, global stiffness matrix creation, material property input, and application of boundary conditions (loading, symmetry, etc.) are the final steps in this phase. In the solution phase, evaluation of engineering variables such as stresses, strain, displacement, and their associated degrees of freedom are computed. The global solution is achieved through the simultaneous solution of various linear or nonlinear equations at the nodes. The last stage referred to as post processing involves the display of results obtained from simulation. These results may include stress, strain, and temperature distributions, deformations, and failure configurations. Most commercially available FEA software requires some critical inputs to make analysis possible, namely, geometry (constructed from nodes), elements, properties, boundary conditions, loading, and type of analysis (for example, static analysis) (Tankut et al. 2014). The main drawbacks associated with FEA are with analysing the huge volume of data generated from simulations and the errors resulting from increasing the number of elements.

The main purpose of this study was to simulate the behaviour of bamboo under various loading conditions and compare the results to real-life data from experimentation.

6.2.2 *Finite Element Analysis of Wood*

The use of FEA models has proven to be advantageous as extrapolation beyond the range of experimental data is possible. While the use of wood products for engineering purposes continues to rise, modelling these structures is more complicated compared to metals such as steel. Several fundamental works utilising FEA and Finite Element Method (FEM) on wood in furniture design have been conducted by some researchers (Eckelman 1967; Eckelman and Suddarth 1969; Smardzewski 1990, 1992; Cai and Wang 1993; Cai et al. 1995; Cai and Chang 1995; Erdil 1995; Gustafsson 1995, 1996a, b, 1997; Smardzewski and Dzięgielewski 1997; Ekström 1997; Chan et al. 1998; Aronsson and Lindgren 2000; Smardzewski and Gawroński 2001; Nicholls and Crisan 2002; Seiichiro and Minoru 2002; Yang et al. 2002; Efe et al. 2003; Pousette 2003; Salokangas 2003; Olsson et al. 2004; Lam et al. 2004; Blanchet et al. 2006; Pousette 2007).

6.2.3 *Finite Element Analysis of Bamboo*

Most research works available (Janssen 1995; Amada and Lakes 1997; Nugroho and Ando 2001; Lo et al. 2004) focus on the characterisation of bamboo through experimentation. However, few works such as those presented by Yu et al. (2005), Silva et al. (2006), Garcia et al. (2012), Khatry and Mishra (2012), and Kingsley et al. (2015) have considered the use of finite element models for estimating the physical and mechanical behaviour of bamboo. In a study by Silva et al. (2006), bamboo was modelled as a functionally graded material, which is a biologically optimised composite. The finite element method and homogenization techniques were employed to capture the variations in material distribution and asymmetric properties, respectively. Tensile, bending, and torsional test simulations were done using a 3D bamboo.

Table 6.1 Material properties from literature

Property	Outer layer (N/mm^2)	Middle layer (N/mm^2)	Inner layer (N/mm^2)	Nodal layer (N/mm^2)
Density	710×10^{-9}	710×10^{-9}	710×10^{-9}	710×10^{-9}
Orthotropic elasticity				
Young modulus X	1.7E+4	1.6E+4	1.2E+4	1.7E+4
Young modulus Y	6.2E+2	4.7E+2	3.4E+2	1.7E+4
Young modulus Z	6.2E+2	4.7E+2	3.4E+2	1.7E+4
Poisson's ratio XY	0.3	0.3	0.3	0.3
Poisson's ratio YZ	0.3	0.3	0.3	0.3
Poisson's ratio XZ	0.3	0.3	0.3	0.3
Shear modulus XY	3.1E+2	2.3E+2	1.7E+2	2.3E+2
Shear modulus YZ	8.5E+3	8E+3	6E+3	8E+3
Shear modulus XZ	3.1E+2	2.3E+2	1.7E+2	2.3E+2
Tensile ultimate strength	2.2E+2	2E+2	1.6E+2	1.06E+2
Compressive ultimate strength	9.20E+1	8.00E+1	6.40E+1	8.00E+1

Adapted from: Khatry and Mishra (2012)

Khatry and Mishra (2012) tabulated bamboo properties from researchers, such as Amada and Untao (2001), Garcia et al. (2012) and Verma et al. (2012), in a format suitable for finite element modelling of orthogonal material as shown in Table 6.1. These values are used as the material input for this section.

6.2.4 Finite Element Characterisation of Bamboo

In this study, the tensile, compressive and bending stresses, and strain properties of bamboo were derived and analysed using finite element analysis

6.2.4.1 Assumptions

To simplify the models, bamboo strips and culms without node were used, while the angle of taper associated with natural bamboo (culm-taper diameter increases towards the ground) was neglected. A non-uniform orthotropic material with linear elastic properties having three equally thick (3 mm each) layers was used.

6.2.4.2 Finite Element Tensile Testing

The element type: Shell 8 node 281.

Figure 6.1 shows the dimensional configuration of the tensile test specimen, while Fig. 6.2 shows the model's three-layered structure. The simulation showed that maximum stress was induced at the mid-section line of the specimen under tension as shown in Fig. 6.3. The maximum Von-Mises strain in Fig. 6.4 is found to act around the neck region of the specimen.

6.2.4.3 Finite Element Compressive Testing

The element type: Solid 8 node 183 (Fig. 6.5).

Material: Internode of inner diameter = 41 mm, Outer diameter = 50 mm and Length = 400 mm.

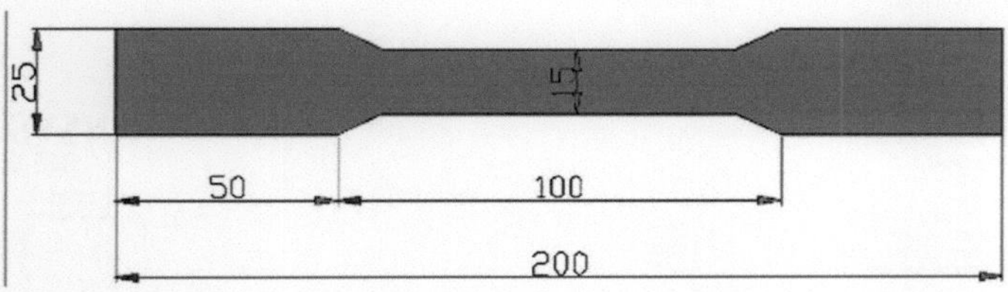

Fig. 6.1 Tensile test specimen

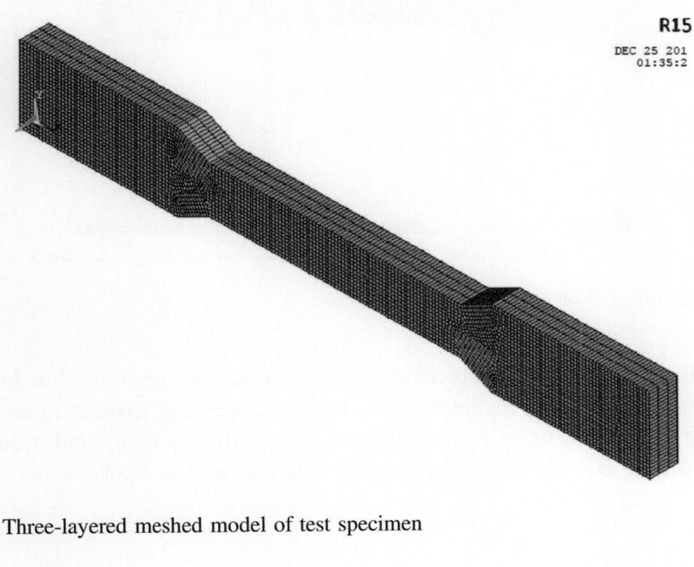

Fig. 6.2 Three-layered meshed model of test specimen

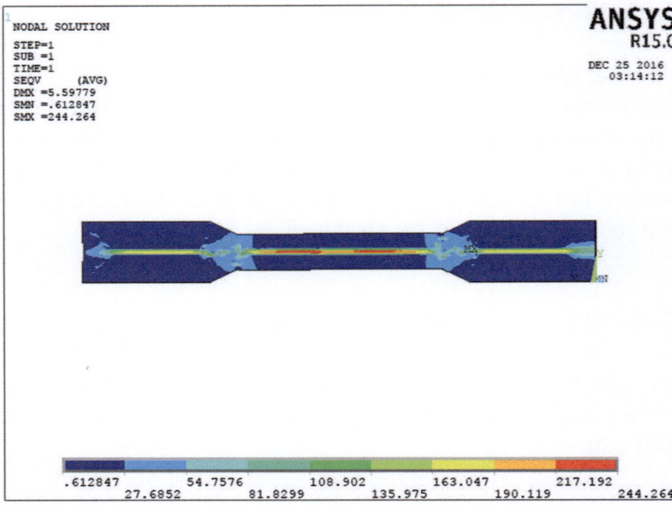

Fig. 6.3 Equivalent (Von-Mises) stress

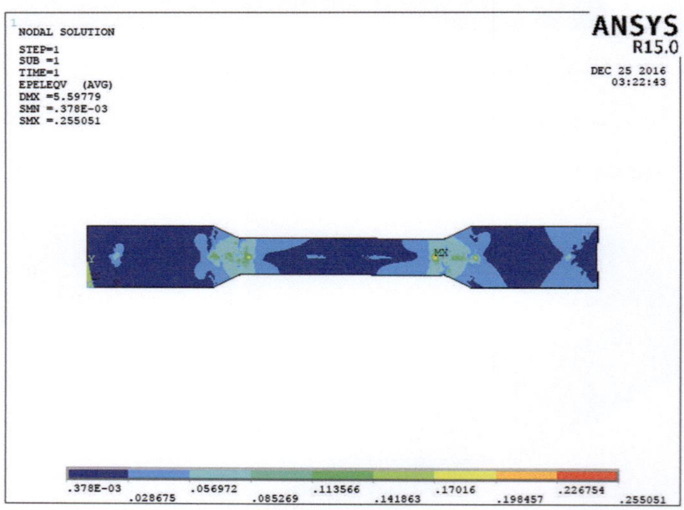

Fig. 6.4 Equivalent (Von-Mises) elastic strain

Discretization of the three-layered culm internode for compression loading is presented in Fig. 6.5. The maximum Von-Mises stresses and maximum shear stress (XY plane) are found to be concentrated at the open ends of the culm that are in contact with the equipment. Failure or crack initiation was likely to occur in this

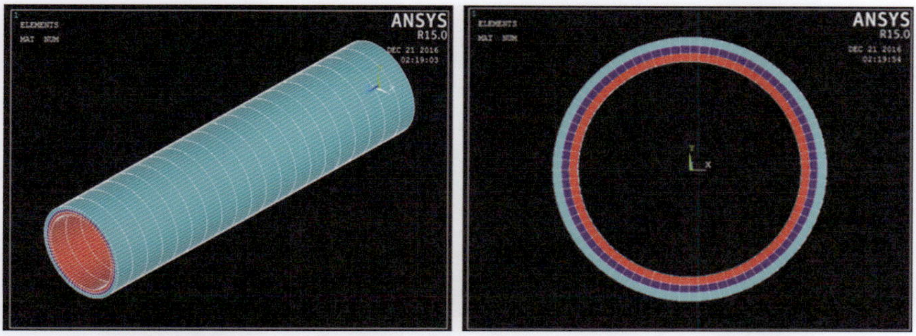

Fig. 6.5 Three (3)-layered meshed model of culm specimen

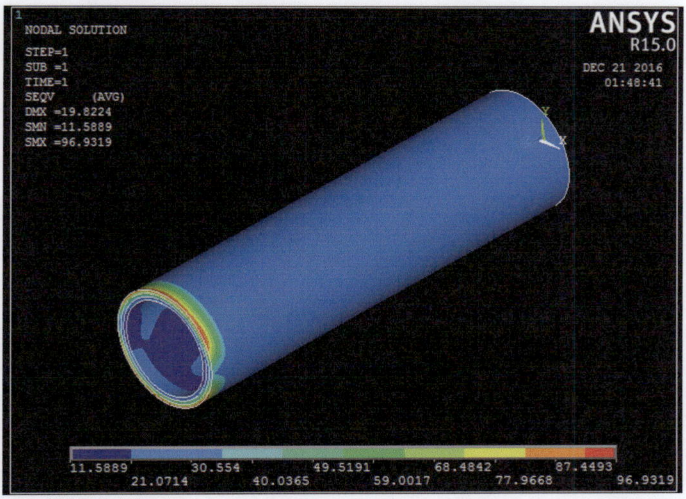

Fig. 6.6 Equivalent (Von-Mises) stress

region of high stress, as shown in Figs. 6.6 and 6.8. The Von-Mises strain, although small, seemed to be spread out through the entire culm as presented in Fig. 6.7.

6.2.4.4 Finite Element Three-Point Bending Testing

The element type: Shell 8 node 281.

For the bending test simulation, symmetry was utilised for better accuracy and simplicity of model as shown in Fig. 6.9, while the three-layered meshed model for FEA analysis is presented in Fig. 6.10. The simulation results, as given in Figs. 6.11 and 6.12, revealed that the induced Von-Mises stress was highest within the inner layer, while the maximum Von-Mises strain was observed in the outer

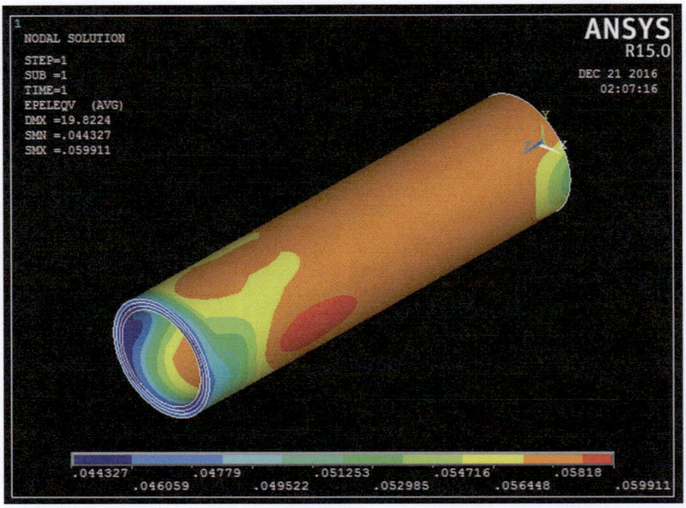

Fig. 6.7 Equivalent (Von-Mises) elastic strain

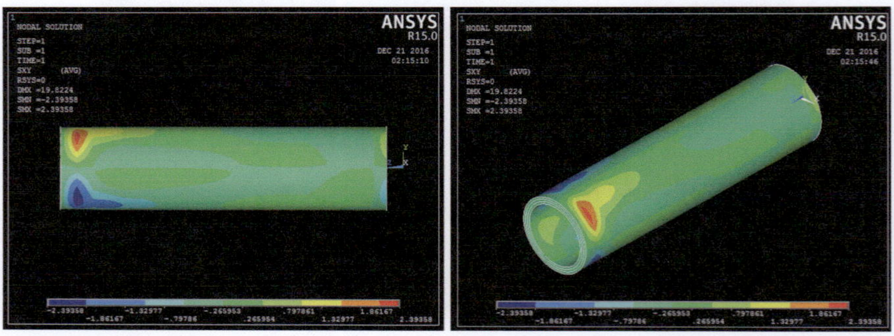

Fig. 6.8 Maximum shear stress XY

layer of the bamboo-culm-wall thickness. The maximum Von-Mises values from the simulation for tensile, compression, and bending testing is presented in Table 6.2.

6.2.5 *Summary*

Finite element analysis was used to successfully predict the behaviour of bamboo under tensile, compressive, and bending loading. This is significant, as bamboo continues to attain more importance in the engineering industry. The ability to

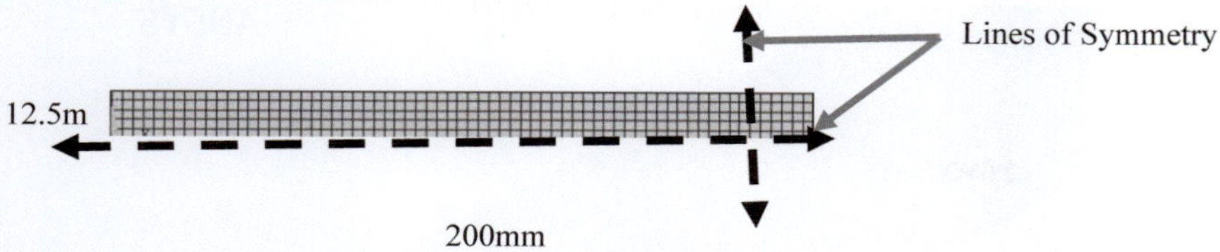

Fig. 6.9 Use of symmetry as a boundary condition

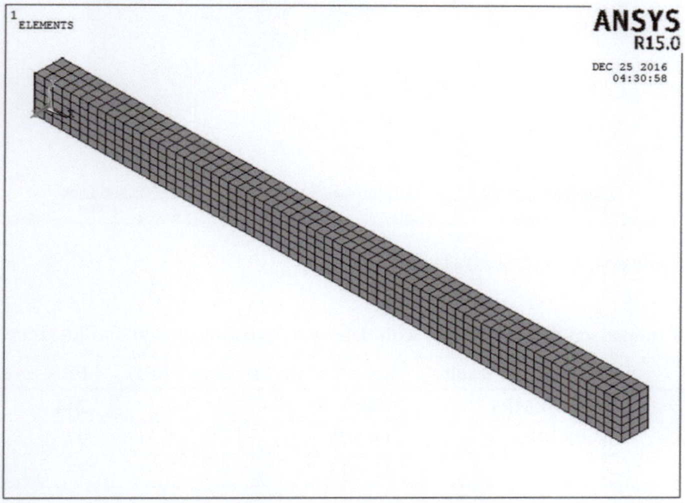

Fig. 6.10 Three-layered meshed model

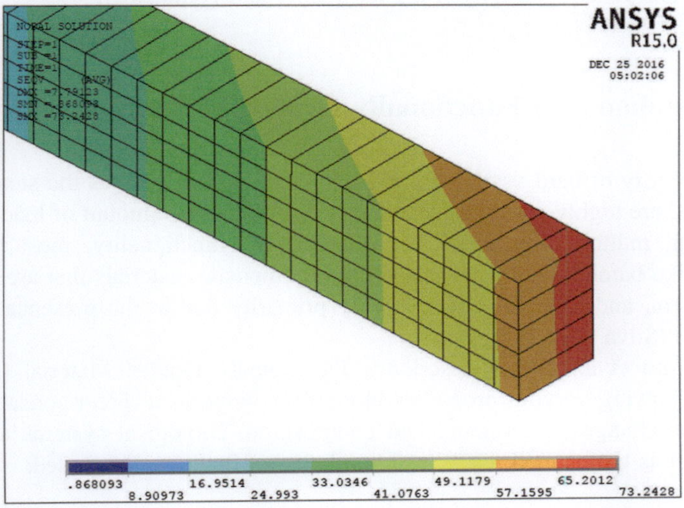

Fig. 6.11 Equivalent (Von-Mises) stress

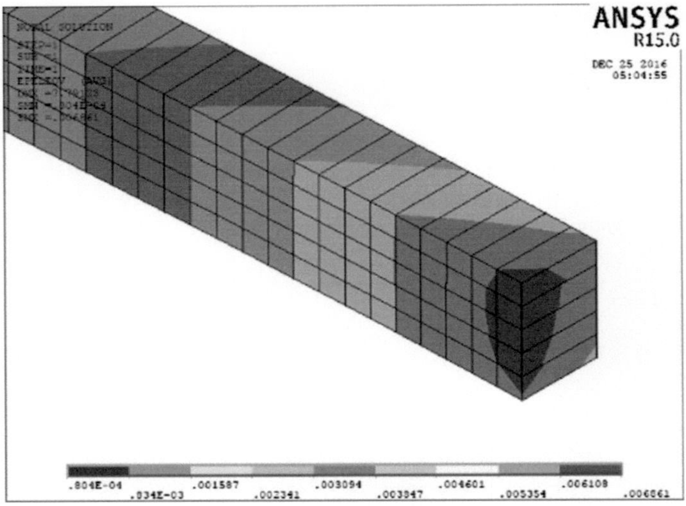

Fig. 6.12 Equivalent (Von-Mises) elastic strain

Table 6.2 Comparison of FEA results with data from experimentation and literature

TEST	Experimental results	Results from literature (MPa)	FEA results (MPa)
Tensile	150–270 MPa	100–315	244
Compressive	88–98 MPa	44–176	97
Bending	–	55–113	73

accurately predict the behaviour of bamboo will lead to much more efficient, effective, and reliable designs of higher quality and complexity.

6.3 Bamboo as a Functionally Graded Material

In the category of hard tissues, many biological entities, such as the stem of trees and bones, are highly optimised to withstand a significant amount of loading while performing multiple functions. To achieve multi-functionality, most biological systems like bamboo culms are made up of composite materials that are complex, non-uniform, and capable of adaptability primarily due to the presence of living organisms (Silva et al. 2006).

Miyamoto et al. (1999) describe a Functionally Graded Material (FGM) as material showing varying properties along its volume as a direct consequence of continuous changes in structure and composition. Biological systems and structures, such as tooth, shell, bone, and bamboo, are functionally graded.

Silva et al. (2008) also defined an FGM as one that has its properties continuously graded. The distribution of the micro-constituent of such materials must be characteristically spatially-varying and non-uniform within all building blocks of the composite. Hence a continuous interchange of composite constituents (matrix and reinforcement) is observed.

Several studies (Nogata and Takahashi 1995; Amada et al. 1996, 1997; Ray et al. 2004) have shown that bamboo stem (culm) may be considered as an FGM. Biological FGMs have the most efficient, economical, and sustainable design required for survival in their highly resourced (energy, minerals, etc.) competitive ecosystems. The ability to adapt geometry (size or shape) to compensate for loading conditions imposed by the environment defines an optimised FGM. Bamboo is a highly optimised FGM that ensures its thriving in varying geographic and environmental conditions.

6.3.1 Bamboo Geometry and Structural Efficiency

Bamboo is a natural composite comprising cellulose fibre and a matrix of lignin. The bamboo culm is made up of 20–30% fibre spread out non-uniformly and longitudinally and can be assumed as an approximately hollow cylindrical shell (Silva et al. 2006). This distribution is observed throughout the entire wall thickness of the culm. Bamboo can, therefore, be considered an orthotropic material because of the fibre distribution and orientation (Nugroho and Ando 2001; Lo et al. 2004). The fibre-distribution concentration decreases progressively from the exterior to the interior of culm wall (Silva et al. 2006).

Bamboo planted on hilly ground showed a unique deformation in the shape of the culm with an almost elliptical or egg-like profile. The deformation may be a result of the nature of loading the culm is being subjected to.

The strength of bamboo radially and axially along the full length was found to be uniform and an indication of optimisation through functional grading, while being structurally efficient with respect to weight cost as presented by Gordon (1978, 1988). From the two studies, pure bamboo fibre exhibited better specific strength when compared with many materials such as alloys and ceramics. Using the rule-of-mixtures, the tensile strength of pure bamboo (810 MPa) was found to be comparable to steel (600–1000 MPa) (Nogata and Takahashi 1995).

6.3.2 Mechanical-Sensing Capabilities

Biological structures such as feline claws and tree stems were thoroughly studied by Mattheck (1990) and Mattheck and Burkhardts (1990), finding that they were optimised structures with respect to their contoured shape and mechanical properties (weight and strength). Nogata and Takahashi (1995) conducted a similar study on bamboo with the use of an electrocardiograph machine to verify if its cells had the capability to sense mechanical loading from the environment. During bending, the compressive side generated higher voltage signals compared to the tension side. It was concluded that, possibly, the voltage signals were an adaptive mechanism used to determine growth processes in a stress-induced environment.

6.3.3 *Shape and Features of the Vascular Bundle*

6.3.3.1 Experimental Work: *Materials and Method*

Bambusa *vulgaris,* which is capable of growing from 15 to 20 m in height, was cut transversely into a section and polished for microstructural examination. Figure 6.13a shows the image from the optical microscope at 2-mm magnification, revealing the distribution of the vascular bundle. The functional graduations can be seen in the different sizes of the bundles. Figure 6.13b, however, is a 200-μm image of the vascular bundle that shows the three main constituents, namely, the xylem which aids in the transfer of water, the phloem which supports the transfer of food to tissues, and the fibre bundles which serve as reinforcement for the phloem and xylem holes, which vary in number (400–500 fibre) depending on the induced-stress distribution (Nogata and Takahashi 1995). Bamboo is regarded as having an optimised structure because the shape and varying number of fibre bundles suggest a compensation for the distribution of stress experienced by the structures in the xylem and phloem. A comparison of the vascular bundle shape

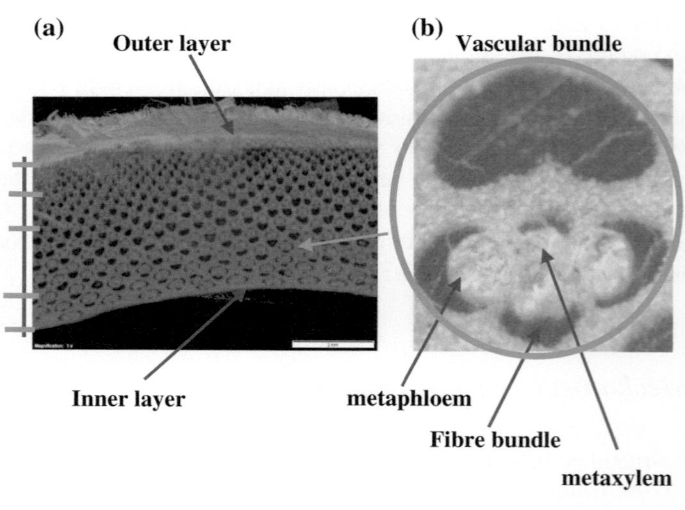

Fig. 6.13 **a** A cross-sectional view of a transverse section of Bambusa Vugaris. **b** An exploded view of the vascular bundle

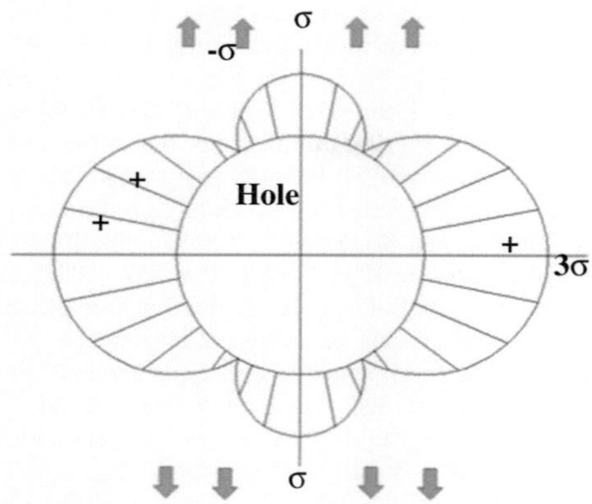

Fig. 6.14 Distribution of stress around a hole embedded within an infinite plate

with the stress distribution induced around a hole within a finite plate (Fig. 6.14) reveals a close similarity (Nogata and Takahashi 1995; Miyamoto et al. 1999).

Nogata and Takahashi (1995) again studied bamboo grown on steep or hilly ground and observed that their vascular bundles were larger and possessed asymmetrical fibre bundles attributable to the type of environmental loading condition.

6.3.4 Summary

Bamboo was shown to be an optimised multiphase FGM that makes it highly adaptable to environmental loading. The experimental work on the microstructure of the vascular bundle conformed to the analyses found in the literature.

References

Amada S (1995) Mater Res Soc Bull 20:35

Amada S, Munekata T, Nagase Y, Ichikawa Y, Kirigai A, Zhifei Y (1996) The mechanical structures of bamboos in viewpoint of functionally gradient and composite materials. J Compos Mater 30:800–819

Amada S, Lakes RS (1997) J Mater Sci 32:2693

Amada S, Ichikawa Y, Munekata T, Nagase Y, Shimizu H (1997) Fiber texture and mechanical graded structure of bamboo. Composites Part B: Engineering 28 (1–2):13–20

Amada S, Untao S (2001) Fracture properties of bamboo. Compos B 32:451–459

Aronsson S, Lindgren CM (2000) Produktutveckling av möbler med hjälp av FEM (Product development of furniture with aid of FEM). Chalmers University of Technology, Göteborg, Sweden (in Swedish)

Arun KB, Ranga RTK and Shankar KS (2002) Matchsticks from Bamboo. J Bamboo and Rattan 2 (4):46–53

Bathe KJ (1996) Finite element procedures. Prentice Hall, New Jersey, USA

Blanchet P, Cloutier A, Gendron G, Beauregard R (2006) Engineered wood flooring design using the finite element method. Forest Prod J. 56(5):65
Cai L, Chang J (1995) Using FEM to analyze heat transfer processes in wood. Holz als Roh-und Werkstoff. ISSN 0018-3768 53(3):183–186. http://dx.doi.org/10.1007/BF02716421
Cai L, Wang F (1993) Influence of the stiffness of corner joints on case furniture deflection. Holz Roh-Werkstoff 51:406–408. doi:10.1007/BF02628238
Cai L, Wang F, Tan H (1995) Study on the strength of moltinject corner joints of furniture. Holz als Roh-und Werkstoff 53(6):385–388. doi:10.1007/s001070050113
Chan SL, Wong KW, So YS, Pon SW (1998) Empirical design and structural performance of bamboo scaffolding. In: Proceedings of the symposium on bamboo and metal Scaffoldings, The Hong Kong Institution of Engineers
Cook RD (1981) Concepts and applications of finite element analysis. Wiley, New York, USA
Eckelman CA (1967) Chair frame analysis and design. For Prod J 17:100–106
Eckelman CA, Suddarth S (1969) Analysis and design of furniture frames. Wood Sci Technol 3:239–255. doi:10.1007/BF00367215
Efe H, Erdil YZ, Kasal A (2003) Optimization of furniture systems with Finite Element Method (FEM) in engineering design of furniture. Technical Education Faculty. Gazi University, Ankara, Turkey
Ekström R (1997) Konstruktionsdesign av möbler med FEM (Design of furniture with FEM). Lund Institute of Technology, Lund, Sweden
Erdil YZ (1995) Integrated product engineering and performance testing of furniture. Ph. D. Thesis, Purdue University, West Lafayette, IN
Garcia JJ, Rangel C, Ghavami K (2012) Experiments with rings to determine the anisotropic elastic constant of bamboo. Constr Build Mater 31:52–57
Ghavami K (1995) Cement & Concrete Compos. 17:281
Gordon JE (1978) Structures or Why Things Don't Fall Down. Penguin Books, London p 17
Gordon JE (1988) The science of structures and materials. Scient. Am. Library, p 176
Gustafsson SI (1995) Furniture design by use of the finite element method. Holz als Roh- und Werkstoff 53(4):257–260. doi:10.1007/s001070050084
Gustafsson SI (1996a) Finite element modelling versus reality for birch chairs. Holz als Roh-und Werkstoff 54(5):355–359. doi:10.1007/s001070050200
Gustafsson SI (1996b) Stability problems in optimised chairs. Wood Sci Technol 39:339–345. doi:10.1007/BF00223553
Gustafsson SI (1997) Optimising ash wood chairs. Wood Sci Technol 3:291–301. doi:10.1007/BF00702616
Janssen JJA (1995) Building with bamboo. Intermediate Technology Publications, London
Khatry R, Mishra DP (2012) Finite element analysis of bamboo column along with steel socket joint under loading condition. Int J Appl Eng Res 7(11)
Kingsley U, Ehi IP, Adgidzi D (2015) Finite element analysis of bamboo bicycle frame. Br J Mathematics & Comput Sci 5(5):583–594
Lam F, Lee G, Yan H, Gu J, Saravi AA (2004) Structural performance of wood-based stair stringers. Forest Prod J. 54(4):39–44
Lo TY, Cui HZ, Leung HC (2004) Mater Lett 58:2595
Mattheck C (1990) Engineering Components grow like trees. Materialwissenschaft und Werkstofftechnik 21(4):143–168
Mattheck C, Burkhardt S (1990) A new method of structural shape optimization based on biological growth. Int J Fatigue 12(3):185–190
Miyamoto Y, Kaysser W, Rabin B, Kawasaki A, Ford R (1999) Functionally graded materials: design, processing and applications. Kluwer Academic Publishers
Nicholls T, Crisan R (2002) Study of the stress-strain state in corner joints and box-type furniture using FEA. Holz als Roh- und Werkstoff 60:66–71. doi:10.1007/s00107-001-0262-0
Nogata F, Takahashi H (1995) Intelligent functionally graded material: Bamboo. Compos Eng 5 (7):743–751
Nugroho N, Ando NJ (2001) Wood Scler 47:237

Olsson P, Eriksson P, Olsson K (2004) Computer-supported furniture design at an early conceptual stage. Int J Des Comput 7

Pousette A (2003) Full-scale test and finite element analysis of a wooden spiral staircase. Holz als Roh-und Werkstoff 61(1):1–7. doi:10.1007/s00107-002-0345-6

Pousette A (2007) Finite element analysis and test of join at the center pole of a wooden spiral stair. Forest Prod J 57(5):25–29

Ray AK, Das SK, Mandal S, Ramachandrarao P (2004) *ibid.* 39:1055

Salokangas L (2003) Wooden observation tower. Helsinki, Finland. Struct Eng Int 3:160–162

Seiichiro U, Minoru M (2002) Wood and wood based materials. Fracture analysis of wood in moment-resisting joints with four drift-pins using Digital Image Correlation method (DIC). J Soc Mat Sci 51(4):367–372. doi:10.1007/s00226-005-0026-9

Silva ECN, Walters MC, Paulino GH (2006) modeling bamboo as a functionally graded material. In: Paulino GH, Pmdera MJ, Dodds RH, Rochinha Jr FA, Dave EV, Chen OL (ed) American Institute of Physics

Smardzewski J (1990) Numerical analysis of furniture construction using finite elements method. Przem Drzewn 41(7):1–5. [In Polish]. doi:10.1007/BF00702895

Smardzewski J (1992) Numeryczna optymalizacja konstrukcji krzesel. Przem Drzewn, No 1

Smardzewski J, Gawroński T (2001) FEM algorithm of chair optimisation. Electron J Pol Agric Univ 4(2)

Smardzewski J., Dzięgielewski S (1997) Computer assisted optimization of selected cross-sections of furniture elements. X Sesja Naukowa Badania dla Meblarstwa. Poznań, pp 9–21

Tankut N, Tankut AN, Zor M (2014) Finite element analysis of wood materials. Drvna Industrija 65(2):159–171

Tanvir WS, Utku M (1987) Sonlu eleman yönteminde varyasyonlu yaklaşım. Orta Doğu Teknik Üniversitesi, Anadolu Üniversitesi, Eskişehir

Verma CS, Chariar VM, Purohit R (2012) Tensile strength analysis of bamboo and layered laminate bamboo composites. Int J Eng Res Appl (IJERA) 2(2):1253–1264

Yang P, Michiyo T, Hayato M, Yasuo O (2002) Structural design improvement of wooden school desks and chairs based on the Hoffman failure criterion. J Jpn Soc Technol Educ 44(3): 133–138

Yu WK, Chung KF, Chan SL (2005) Axial Buckling of bamboo columns in a bamboo scaffolds. Eng Struct 27:61–73

Zienkiewicz OC, Taylor RL (1988) The finite element method: vol. 1-basic formulation and linear problems. McGraw Hill, London

Zienkiewicz, OC, Taylor RL (1989) The finite element method: vol. 2-solid and fluid mechanics, dynamics and non-linearity. McGraw Hill, London

Appendix

Classification of Bamboo

Genus	Species	Origin
Acidosasa		**China**
	Acidosasa chienouensis	
	Acidosasa edulis	
	Acidosasa notate	
Ampelocalamus		**China and Nepal**
Arundinaria		**USA**
	Arundinaria appalachiana	
	Arundinaria funghomii	
	Arundinaria gigantea	
	Arundinaria gigantea 'Macon'	
	Arundinaria tecta	
	Bambusa arnhemica	Australia
	Bambusa arnhemica	Australia
	Bambusa balcooa	India, Nepal
	Bambusa bambos	India
	Bambusa basihirsuta	
	Bambusa beecheyana	
	Bambusa beecheyana var pubescens	
	Bambusa blumeana	
	Bambusa boniopsis	
	Bambusa burmanica	
	Bambusa chungii	
	Bambusa chungii var. barbelatta	
	Bambusa cornigera	
	Bambusa dissimulator	
	Bambusa dissimulator 'Albinodia'	
	Bambusa distegia	China
	Bambusa dolichoclada	

(continued)

E.T. Akinlabi et al., *Bamboo*, DOI 10.1007/978-3-319-56808-9

(continued)

Genus	Species	Origin
Bambusa		
	Bambusa dolichoclada 'Stripe'	Taiwan
	Bambusa dolichomerithalla 'Green stripe'	Taiwan
Bambusa		
	Bambusa dolichomerithalla 'Silverstripe'	Taiwan
	Bambusa emeiensis	
	Bambusa emeiensis 'Chrysotrichus'	
	Bambusa emeiensis 'Flavidovirens'	
	Bambusa emeiensis 'Viridiflavus'	
	Bambusa eutuldoides	Hong Kong
	Bambusa eutuldoides 'Viridivittata'	
	Bambusa flexosa	
	Bambusa fulda	
	Bambusa gibba	
	Bambusa glaucophylla	
	Bambusa intermedia	
	Bambusa lako	China
	Bambusa longispiculata	
	Bambusa maculate	
	Bambusa multiplex	
	Bambusa multiplex 'Alphonse Karr'	
	Bambusa multiplex 'Fernleaf Stripestem'	
	Bambusa multiplex 'Golden Goddess'	
	Bambusa multiplex 'Goldstripe'	
	Bambusa multiplex 'Lucky Silverstripe'	
	Bambusa multiplex 'Riviereorum'	
	Bambusa multiplex 'Silver stripe'	
	Bambusa multiplex 'Tiny Fern Striped'	
	Bambusa multiplex 'Tiny Fern'	
	Bambusa multiplex 'Willowy'	

(continued)

(continued)

Genus	Species	Origin
	Bambusa nutans	Thailand
	Bambusa odashimae	Taiwan
	Bambusa odashimae X B. tuldoides	
	Bambusa oldhamii	U.S. A
	Bambusa oliveriana	
	Bambusa pachinensis	
	Bambusa pervariabilis	China
	Bambusa pervariabilis 'Vi ridistriatus'	China
	Bambusa rigida	
	Bambusa rutile	
	Bambusa sinospinosa	
	Bambusa sp. 'Hirose'	
	Bambusa sp. 'Clone X'	
	Bambusa sp. 'Nana'	Thailand
	Bambusa sp. 'Polymorpha'	Nepal
	Bambusa sp. 'Richard Waldron'	
	Bambusa suberecta	
	Bambusa textilis	
	Bambusa textilis 'Dwarf'	
	Bambusa textilis 'Kanapaha'	
	Bambusa textilis 'Maculata	
	Bambusa textilis 'Mutabilis'	
	Bambusa textilis 'Scranton'	
	Bambusa textilis var. al bostriata	
	Bambusa textilis var. glabra	
	Bambusa textilis var. gracilis	
	Bambusa oldhamii	U.S. A
	Bambusa oliveriana	
	Bambusa pachinensis	
	Bambusa pervariabilis	China
	Bambusa pervariabilis 'Viridistriatus'	China
	Bambusa rigida	
	Bambusa rutile	
	Bambusa sinospinosa	
	Bambusa sp 'Hirose'	
	Bambusa sp. 'Clone X'	
	Bambusa sp. 'Nana'	Thailand
	Bambusa sp. 'Polymorpha'	Nepal

(continued)

(continued)

Genus	Species	Origin
	Bambusa sp. 'Richard Waldron'	
	Bambusa suberecta	
	Bambusa textilis	
	Bambusa textilis 'Dwarf'	
	Bambusa textilis 'Kanapaha'	
	Bambusa textilis 'Maculata	
	Bambusa textilis 'Mutabilis'	
	Bambusa textilis 'Scranton'	
	Bambusa textilis var. albostriata	
	Bambusa textilis var. glabra	
	Bambusa textilis var. gracilis	
	Bambusa tulda	
	Bambusa tulda 'Striata'	
	Bambusa tuldoides	
	Bambusa variostriata	
	Bambusa ventricosa	
	'Golden Buddha'	
	Bambusa ventricosa 'Kimmei'	
	Bambusa vulgaris	Tropics
	Bambusa vulgaris 'Vittata'	
	Bambusa vulgaris 'Wamin Striata'	
	Bambusa vulgaris 'Wamin'	
Bashania		**China**
	Bashania fargesii	China
	Bashania qingchengshanensis	
Bergbambos		**Southern Africa**
	Bergbambos tessellate	
Borinda		**Bhutan, China, Nepal, Yunnan**
	Borinda KR 5288	China
	Borinda albocerea *Borinda angustissima* *Borinda contracta* *Borinda frigidorum* *Borinda fungosa* *Borinda fungosa* 'White Cloud' *Borinda grossa* *Borinda lushuiensis* *Borinda macclureana* *Borinda nujiangensis* *Borinda papyrifera*	

(continued)

(continued)

Genus	Species	Origin
	Borinda perlonga *Borinda* sp. 'Muliensis'	
	Borinda yulongshanensis	
Brachystachyum		**China**
	Brachystachyum densiflorum	
	Brachystachyum densiflorum var. villosum	
Cephalostachyum		**India, China, Indonesia, Madagascar, Malaysia**
	Cephalostachyum pergracile	
	Cephalostachyum virgatum	
Chimonobambusa		**China**
	Chimonobambusa macrophylla 'Intermedia'	
	Chimonobambusa makuanensis	
	Chimonobambusa marmoreal	China
	Chimonobambusa marmorea 'Variegata'	
	Chimonobambusa pachystachys	
	Chimonobambusa quadrangularis	
	Chimonobambusa quadrangularis 'Joseph de Jussieu'	
	Chimonobambusa quadrangularis 'Suow'	
	Chimonobambusa quadrangularis 'Yellow Groove'	
	Chimonobambusa tumidissinoda	China
	Chimonobambusa yunnanensis	
Chimonocalamus		**China**
	Chimonocalamus pallens	
Chusquea		**Central and South America, Chile, Argentina, Mexico, Costa Rica, Brazil**
	Chusquea andina	
	Chusquea andina 'Blue Andes'	Chile
	Chusquea bilimekii	Mexico
	Chusquea circinata	Mexico
	Chusquea circinata 'Chiapas'	
	Chusquea coronalis	
	Chusquea culeou	Chile

(continued)

(continued)

Genus	Species	Origin
Chusquea		**Central and South America, Chile, Argentina, Mexico, Costa Rica, Brazil**
	Chusquea culeou 'Argentina'	Argentina
	Chusquea culeou 'Hillier's Form'	
	Chusquea culeou 'Llaima'	
	Chusquea cumingii	
	Chusquea delicatula	
	Chusquea foliosa	
	Chusquea galeottiana	Chile
	Chusquea gigantean	Mexico
	Chusquea glauca	Mexico
	Chusquea liebmannii	Costa Rica and Mexico
	Chusquea mimosa ssp. Australis	Brazil
	Chusquea montana	Chile
	Chusquea muelleri	Mexico
	Chusquea pittieri	
	Chusquea simpliciflora	
	Chusquea sp. 'Chiconquiaco'	Mexico
	Chusquea sp. 'Las Vigas'	
	Chusquea subtilis	Costa Rica
	Chusquea sulcata	Mexico and Costa Rica
	Chusquea uliginosa	Chile
	Chusquea valdiviensis	Chile
	Chusquea virgate	Costa Rica
Dendrocalamus		**Tropical regions, China and Vietnam**
	Dendrocalamus asper	
	Dendrocalamus asper 'Betung Hitam'	
	Dendrocalamus brandisii	
	Dendrocalamus brandisii 'Black'	
	Dendrocalamus brandisii (variegated)	
	Dendrocalamus calostachyus	
	Dendrocalamus giganteus	
	Dendrocalamusgiganteus (Quail Clone)	

(continued)

(continued)

Genus	Species	Origin
	Dendrocalamus giganteus (variegated)	
	Dendrocalamus grandis	
	Dendrocalamus hamiltonii	
	Dendrocalamus jianshuiensis	China
	Dendrocalamus ianshuiensis (variegated)	
	Dendrocalamus latiflorus 'Mei-nung'	
	Dendrocalamus membranaceus	
	Dendrocalamus minor	
	Dendrocalamus minor 'Amoenus'	
	Dendrocalamus sikkimensis	
	Dendrocalamus sp. 'Maroochy'	China
	Dendrocalamus sp. 'Parker's Giant' *Dendrocalamus strictus*	
	Dendrocalamus validus	
	Dendrocalamus yunnanicus	China and Vietnam
Dinochloa		**Philippines, Burma, Malaysia, Thailand**
	Dinochloa malayana	Malaysia and Thailand
	Dinochloa scandens	
Drepanostachyum		**India, Nepal, Bhutan, and China**
	Drepanostachyum falcatum var. sengteeanum	
	Drepanostachyum fractiflexum	
Fargesia		**China**
	Fargesia Sp. 'Tabashan II'	
	Fargesia adpressa	
	Fargesia apicirubens	China
	Fargesia apicirubens 'White Dragon'	
	Fargesia communis	
	Fargesia denudata	
	Fargesia denudata 'Xian 1'	
	Fargesia denudata 'Xian 2'	
	Fargesia dracocephala 'Rufa'	
	Fargesia gaolinensis	China
	Fargesia murielae 'Bimbo'	

(continued)

(continued)

Genus	Species	Origin
	Fargesia murielae 'Harewood'	
	Fargesia murielae 'Jonny's Giant'	
	Fargesia murielae 'Simba'	
	Fargesia murieliae	China
	Fargesia murieliae 'SABE 939'	
	Fargesia murieliae 'Vampire'	
	Fargesia nitida	
	Fargesia nitida 'Anceps'	
	Fargesia nitida 'Gansu 2'	
	Fargesia nitida 'Jiuzhaigou 10'	
	Fargesia nitida 'Jiuzhaigou 4'	China
	Fargesia nitida 'Jiuzhaigou Genf'	
	Fargesia nitida 'Jiuzhaigou'	
	Fargesia nitida 'Nymphenburg' *Fargesia nitida* 'Wakehurst' *Fargesia robusta*	China
Gaoligongshania		**Indonesia**
	Gaoligongshania megalothyrsa	
	Gigantochloa Hitam Hijau	
	Gigantochloa apus	
	Gigantochloa atroviolacea	
	Gigantochloa atroviolacea 'Watupawan'	
	Gigantochloa albociliata	
	Gigantochloa atter	
	Gigantochloa hasskarliana	Indonesia
	Gigantochloa levis	
	Gigantochloa ligulata	
	Gigantochloa luteostriata	
	Gigantochloa maxima	
	Gigantochloapseudoarundinacea	
	Gigantochloa ridleyi	
	Gigantochloa robusta	
	Gigantochloa sp 'Rachel Carson'	
	Gigantochloa sp. 'Bali White Stripe'	
	Gigantochloa sp. 'Sumatra 3751'	
	Gigantochloa sp. 'Widjaja 3827'	
	Gigantochloa sp. Marga	
	Gigantochloa wrayii	Malaysia

(continued)

(continued)

Genus	Species	Origin
Guadua		**Peru, Venezuela, Bolivia, Uruguay, USA**
	Guadua amplexifolia	
	Guadua angustifolia	Peru and Venezuela
	Guadua angustifolia 'Bicolor'	
	Guadua angustifolia 'Less Thorny'	
	Guadua chacoensis	Bolivia and Uruguay
	Guadua longifolia	
	Guadua paniculata	USA
	Guadua sp. 'Aureocaulis'	
	Guadua velutina	
Hibanobambusa		**Japan**
	Hibanobambusa tranquillans	
	Hibanobambusa tranquillans 'Shiroshima'	
Himalayacalamus		**Bhutan, India, Nepal, China**
	Himalayacalamus cupreus *Himalayacalamus falconeri*	
	Himalayacalamus falconeri 'Damarapa'	
	Himalayacalamus fimbriatus	
	Himalayacalamus hookerianus	
	Himalayacalamus intermedius	
	Himalayacalamus planatus	Nepal
	Himalayacalamus porcatus	Nepal
Indocalamus		**China**
	Indocalamus cordatus	
	Indocalamus decorus	
	Indocalamus latifolius	
	Indocalamus latifolius 'Hopei'	
	Indocalamus longiauritus	
	Indocalamus migoi	
	Indocalamus pedalis *Indocalamus* sp. 'Hamadae' *Indocalamus* sp. 'Solidus' *Indocalamus tessellatus* *Indocalamus victorialis*	Japan
Indosasa		**China and Vietnam**
	Indosasa crassiflora *Indosasa gigantea*	China
	Indosasa ingens	China

(continued)

(continued)

Genus	Species	Origin
Lithachne		**Tropical Western Hemisphere**
	Lithachne humilis	
Melocalamus		**China**
	Melocalamus arrectus	China
Melocanna		**India, Burma and USA**
	Melocanna baccifera	
Menstruocalamus		**China**
	Menstruocalamus sichuanensis	China
Nastus		**Madagascar and Solomon Islands**
	Nastus elatus	Papua New Guinea
Neohouzeaua		**South East Asia**
	Neohouzeaua mekongensis	Vietnam
Neololeba		**Austria, Papua New Guinea and Indonesia**
	Neololeba atra	Austria
Ochlandra		**India and Sri Lanka**
	Ochlandra stridula	Sri Lanka
Oldeania		**Central Africa**
	Oldeania alpina	Africa
Oligostachyum		China
	Oligostachyum glabrescens	China
Olmeca		Mexico
	Olmeca recta	
Otatea		**Central Mexico and Central America**
	Otatea acuminate 'Michoacan'	
	Otatea acuminata ssp. Acuminata	Eastern Mexico
	Otatea acuminata ssp. Aztecorum	
	Otatea acuminata ssp. aztecorum 'Dwarf'	
	Otatea fimbriata	Mexico and Colombia
	Otatea glauca 'Mayan Silver'	Mexico
Oxytenanthera		**Savanna woodlands of tropical Tropical Africa**
	Oxytenanthera abyssinica	Ethiopia
	Oxytenanthera braunii	Tanzania

(continued)

(continued)

Genus	Species	Origin
Phyllostachys		**China, India, USA**
	Phyllostachys acuta	China
	Phyllostachys angusta	China
	Phyllostachys arcana	China
	Phyllostachys arcana 'Luteosulcata'	
	Phyllostachys atrovaginata	
	Phyllostachys aurea	USA
	Phyllostachys aurea 'Albovariegata'	
	Phyllostachys aurea 'Holochrysa'	
	Phyllostachys aurea 'Koi'	
	Phyllostachys aurea 'Takemurai'	
	Phyllostachys aureosulcata *Phyllostachys aureosulcata* 'Alata'	
	Phyllostachys aureosulcata 'Aureocaulis'	
	Phyllostachys aureosulcata 'Harbin Inversa'	
	Phyllostachys aureosulcata 'Harbin'	
	Phyllostachys aureosulcata 'Lama Temple'	
	Phyllostachys aureosulcata 'Pekinensis'	
	Phyllostachys aureosulcata 'Spectabilis'	
	Phyllostachys aurita	
	Phyllostachys bambusoides	Japan
	Phyllostachys bambusoides 'Albovariegata'	
	Phyllostachys bambusoides 'Allgold'	
	Phyllostachys bambusoides 'Aureostriata'	Japan
	Phyllostachys bambusoides 'Castillon Inversa'	
	Phyllostachys bambusoides 'Castillon variegata'	

(continued)

(continued)

Genus	Species	Origin
Phyllostachys		**China, India, USA**
	Phyllostachys bambusoides 'Castillon'	
	Phyllostachys bambusoides 'Golden Dwarf'	
	Phyllostachys bambusoides 'Job's Spots'	
	Phyllostachys bambusoides 'Kawadana'	
	Phyllostachys bambusoides 'Leprechaun Gold'	
	Phyllostachys bambusoides 'Marliac'	
	Phyllostachys bambusoides 'Rib Leaf'	
	Phyllostachys bambusoides 'Richard Haubrich'	
	Phyllostachys bambusoides 'Slender Crookstem'	
	Phyllostachys bambusoides 'Subvariegata'	
	Phyllostachys bambusoides 'Tanakae'	
	Phyllostachys bambusoides 'White Crookstem'	
	Phyllostachys bambusoides 'katashibo'	
	Phyllostachys bissetii	
	Phyllostachys bissetii 'Dwarf'	
	Phyllostachys dulcis	
	Phyllostachys edulis	
	Phyllostachys edulis 'Anderson'	
	Phyllostachys edulis 'Bicolor'	
	Phyllostachys edulis 'Goldstripe'	
	Phyllostachys edulis 'Gracilis'	
	Phyllostachys edulis 'Heterocycla'	Japan
	Phyllostachys edulis 'Jaquith'	
	Phyllostachys edulis 'Nabeshimana'	
	Phyllostachys edulis 'Tubaeformis'	China
	Phyllostachys fimbriligula	China

(continued)

(continued)

Genus	Species	Origin
	Phyllostachys flexuosa	
	Phyllostachys flexuosa 'Kimmei Aureostriata'	
	Phyllostachys flexuosa 'Kimmei'	
	Phyllostachys glabrate	
	Phyllostachys glauca	China
	Phyllostachys glauca 'Notso'	
	Phyllostachys glauca 'Yunzhu'	
	Phyllostachys heteroclada	
	Phyllostachys heteroclada 'Purpurata'	
	Phyllostachys heteroclada 'Solidstem'	
	Phyllostachys hispida	Japan
	Phyllostachys humilis	
Phyllostachys		**China, India, USA**
	Phyllostachys incarnata	
	Phyllostachys iridescens	
	Phyllostachys kwangsiensis	
	Phyllostachys lithophila	
	Phyllostachys lofushanensis	
	Phyllostachys makinoi	China
	Phyllostachys mannii 'Decora'	
	Phyllostachys mannii 'Mannii'	China and India
	Phyllostachys meyeri	
	Phyllostachys nidularia	
	Phyllostachys nidularia 'Farcta'	
	Phyllostachys nidularia 'Smoothsheath'	
	Phyllostachys nidularia June Barbara	
	Phyllostachys nigra	
	Phyllostachys nigra 'Bory'	
	Phyllostachys nigra 'Daikokuchiku'	
	Phyllostachys nigra 'Hale'	
	Phyllostachys nigra 'Henon Dwarf'	
	Phyllostachys nigra 'Henon'	
	Phyllostachys nigra 'Megurochiku'	

(continued)

(continued)

Genus	Species	Origin
	Phyllostachys nigra 'Mejiro'	
	Phyllostachys nigra 'Muchisasa'	
	Phyllostachys nigra 'Othello'	
	Phyllostachys nigra 'Punctata'	
	Phyllostachys nigra 'Shimadake'	
	Phyllostachys nigra 'Tosaensis'	
	Phyllostachys nuda	
	Phyllostachys nuda 'Localis'	
	Phyllostachys parvifolia	
	Phyllostachys pingyangensis	
	Phyllostachys platyglossa	China
	Phyllostachys praecox	China
	Phyllostachys praecox 'Prevernalis'	
	Phyllostachys praecox 'Viridisulcata'	
	Phyllostachys prominens	
	Phyllostachys propinqua	China
	Phyllostachys propinqua 'Beijing'	
Phyllostachys		**China, India, USA**
	Phyllostachys rigida	
	Phyllostachys robustiramea	
	Phyllostachys rubromarginata	
	Phyllostachys rutila	
	Phyllostachys stimulosa	China
	Phyllostachys varioauriculata	
	Phyllostachys violascens	
	Phyllostachys virella	
	Phyllostachys viridiglaucescens	
	Phyllostachys viridis	
	Phyllostachys viridis 'Houzeau'	
	Phyllostachys viridis 'Robert Young'	
	Phyllostachys vivax *Phyllostachys vivax* 'Aureocaulis' *Phyllostachys vivax* 'Black Spot'	
	Phyllostachys vivax 'Huangwenzhu Inversa'	
	Phyllostachys ivax 'Huangwenzhu'	

(continued)

(continued)

Genus	Species	Origin
Pleioblastus		**Japan, China**
	Pleioblastus akebono	
	Pleioblastus amarus	China
	Pleioblastus argenteostriatus	
	Pleioblastus argenteostriatus 'Goldstripe'	
	Pleioblastus chino	Japan
	Pleioblastus chino 'Angustifolia'	
	Pleioblastus chino 'Elegantissimus'	
	Pleioblastus chino 'Gracilis'	
	Pleioblastus chino 'Kimmei'	
	Pleioblastus chino 'Murakamiansus'	
	Pleioblastus chino 'Vaginatus Variegatus'	
	Pleioblastus distichus	
	Pleioblastus distichus 'Mini'	
	Pleioblastus fortunei	
	Pleioblastus gauntlettii	
	Pleioblastus gozadakensis	
	Pleioblastus gramineus	
	Pleioblastus gramineus 'Monstrispiralis'	
	Pleioblastus hindsii	
Pleioblastus		**China, India, USA**
	Pleioblastus humilis	
	Pleioblastus humilis 'Albovariegatus'	
	Pleioblastus humilis 'Variegatus'	
	Pleioblastus juxianensis	China
	Pleioblastus kodzumae	
	Pleioblastus kongosanensis	
	Pleioblastus kongosanensis 'Akibensis'	
	Pleioblastus kongosanensis 'Aureostriatus'	
	Pleioblastus linearis	Taiwan
	Pleioblastus linearis 'Nana'	
	Pleioblastus maculatus	
	Pleioblastus nagashima	Japan

(continued)

(continued)

Genus	Species	Origin
	Pleioblastus oleosus	
	Pleioblastus pygmaeus	
	Pleioblastus pygmaeus 'Greenstripe'	
	Pleioblastus pygmaeus 'Ramosissimus'	
	Pleioblastus shibuyanus 'Tsuboi'	
	Pleioblastus simonii	
	Pleioblastus simonii 'Variegatus'	
	Pleioblastus viridistriatus	
	Pleioblastus viridistriatus 'Chrysophyllus'	
	Pleioblastus xestrophyllus	
Pseudosasa		**China, Japan, Taiwan**
	Pseudosasa amabilis	China
	Pseudosasa cantori	China
	Pseudosasa guanxianensis	China
	Pseudosasa japonica	
	Pseudosasa japonica 'Akebono'	
	Pseudosasa japonica 'Akebono-suji'	
	Pseudosasa japonica 'Pleioblastoides'	
	Pseudosasa japonica 'Tsutsumiana'	
	Pseudosasa japonica 'Variegata'	
	Pseudosasa longiligula	China
	Pseudosasa owatarii	Japan
	Pseudosasa usawai	Taiwan
	Pseudosasa viridula	
Raddia		
	Raddia brasiliensis	
	Raddia distichophylla	
Rhipidocladum		**Argentina and Mexico, Coata Rica**
	Rhipidocladum pittieri	Costa Rica
	Rhipidocladum racemiflorum	
Sarocalamus		**Bhutan, India, Nepal, China**
	Sarocalamus faberi	
	Sarocalamus fangianus	China

(continued)

(continued)

Genus	Species	Origin
Sasa		**Japan**
	Sasa cernua	
	Sasa gracillima	
	Sasa hayatae	
	Sasa kagamiana	
	Sasa kagamiana ssp. Yoshinoi	
	Sasa kurilensis	
	Sasa kurilensis 'Shimofuri'	
	Sasa megalophylla	Japan
	Sasa nagimontana	Japan
	Sasa nipponica (hort.)	Japan
	Sasa oshidensis	
Sasa		**Japan**
	Sasa palmata	
	Sasa senanensis	
	Sasa shimidzuana	
	Sasa sp.	
	Sasa tsuboiana	Japan
	Sasa veitchii	
	Sasa veitchii 'Borealis striata'	
Sasaella		**Japan**
	Sasaella bitchuensis *Sasaella hidaensis* 'muraii'	Japan Japan
	Sasaella masamuneana	
	Sasaella masamuneana 'Albostriata'	
	Sasaella masamuneana 'Aureostriata'	
	Sasaella ramosa	
	Sasaella sasakiana	
	Sasaella shiobarensis	Japan
Sasamorpha		**Japan**
	Sasamorpha borealis	Japan
Schizostachyum		**Asia regions, Indonesia, Solomon Islands**
	Schizostachyum brachycladum	Asia
	Schizostachyum brachycladum 'Bali Kuning'	
	Schizostachyum caudatum	Indonesia
	Schizostachyum glaucifolium	
	Schizostachyum jaculans	

(continued)

(continued)

Genus	Species	Origin
	Schizostachyum lima	
	Schizostachyum sp. 'Murray Island'	Solomon Islands
	Schizostachyum zollingeri	
Semiarundinaria		**Japan, Korea**
	Semiarundinaria fastuosa	
	Semiarundinaria fastuosa 'Viridis'	
	Semiarundinaria fortis	Japan
	Semiarundinaria kagamiana	
	Semiarundinaria lubrica	
	Semiarundinaria makinoi	
Semiarundinaria		**Japan, Korea**
	Semiarundinaria okuboi	
	Semiarundinaria sp. Maruyamana	
	Semiarundinaria sp. 'Korea'	Korea
	Semiarundinaria yamadori	
	Semiarundinaria yamadori 'Brimscomb'	
	Semiarundinaria yashadake	
	Semiarundinaria yashadake 'Kimmei'	
	Semiarundinaria yashadake 'kimmei inversa'	
Shibataea		
	Shibataea chinensis	
	Shibataea kumasaca	
	Shibataea kumasaca 'Albostriata'	
	Shibataea kumasaca 'Aureostriata'	
	Shibataea lancifolia	
	Shibataea nanpingensis	
Sinobambusa		
	Sinobambusa gigantea	
	Sinobambusa intermedia	
	Sinobambusa tootsik	
	Sinobambusa tootsik 'Albostriata'	

(continued)

(continued)

Genus	Species	Origin
Thamnocalamus		**Africa, Eastern Asia regions, Nepal**
	Thamnocalamus crassinodus	
	Thamnocalamus crassinodus 'Aristatus hort. US'	Nepal
	Thamnocalamus crassinodus 'Kew Beauty'	
	Thamnocalamus crassinodus 'Mendocino'	
	Thamnocalamus crassinodus 'Merlyn'	
	Thamnocalamus nepalensis 'Nyalam'	
Thamnocalamus	*Thamnocalamus spathiflorus*	
	Thamnocalamus spathiflorus aff. 'Shivapuri'	Eastern Himalayas
Thyrsostachys		**Tropical regions, Thailand**
	Thyrsostachys oliveri	Tropical
	Thyrsostachys siamensis	Thailand
Vietnamosasa		**Cambodia, Vietnam, Laos, and Thailand**
	Vietnamosasa ciliata	
	Vietnamosasa pusilla	
Yushania		**Africa, India and Taiwan, Nepal, China, Japan**
	Yushania anceps	India
	Yushania anceps 'Pitt White'	
	Yushania boliana	Japan
	Yushania brevipaniculata	China
	Yushania brevipaniculata 'Wolong'	
	Yushania exilis	
	Yushania lu	Himalaya
	Yushania maculata	Sichuan and Yunnan
	Yushania maling	Nepal

Adapted from Guadua Bamboo Company (2016, November 20)

Index

E.T. Akinlabi et al., *Bamboo*, DOI 10.1007/978-3-319-56808-9

MIX
Papier aus verantwortungsvollen Quellen
Paper from responsible sources
FSC® C105338

If you have any concerns about our products,
you can contact us on
ProductSafety@springernature.com

In case Publisher is established outside the EU,
the EU authorized representative is:
Springer Nature Customer Service Center GmbH
Europaplatz 3, 69115 Heidelberg, Germany

Printed by Libri Plureos GmbH
in Hamburg, Germany